**Harcourt**

Orlando   Austin   Chicago   New York   Toronto   London   San Diego

Visit *The Learning Site!*
www.harcourtschool.com

Copyright © by Harcourt, Inc.
2007 Edition

All rights reserved. No part of this publication may be reproduced or transmitted in any form or by any means, electronic or mechanical, including photocopy, recording, or any information storage and retrieval system, without permission in writing from the publisher.

Requests for permission to make copies of any part of the work should be addressed to School Permissions and Copyrights, Harcourt, Inc., 6277 Sea Harbor Drive, Orlando, Florida 32887-6777. Fax: 407-345-2418.

HARCOURT and the Harcourt Logo are trademarks of Harcourt, Inc., registered in the United States of America and/or other jurisdictions.

For permission to reprint copyrighted material, grateful acknowledgment is made to the following sources:

Candlewick Press Inc., Cambridge, MA: Cover illustration by Nick Sharratt from *Isn't It Time?* by Judy Hindley. Illustration copyright © 1994 by Nick Sharratt. Cover illustration by Cynthia Jabar from *How Many, How Many, How Many* by Rick Walton. Illustration copyright © 1993 by Cynthia Jabar.

Harcourt, Inc.: Cover illustration from *The Great Kapok Tree: A Tale of the Amazon Rain Forest* by Lynne Cherry. Copyright © 1990 by Lynne Cherry. Cover illustration from *Pancakes for Breakfast* by Tomie dePaola. Copyright © 1978 by Tomie dePaola. Cover illustration from *Fish Eyes* by Lois Ehlert. Copyright © 1990 by Lois Ehlert. Cover illustration by Pamela Lofts from *Koala Lou* by Mem Fox. Illustration copyright © 1988 by Pamela Lofts.

Philomel Books, an imprint of Penguin Books for Young Readers, a division of Penguin Putnam Inc.: Cover illustration from *The Very Busy Spider* by Eric Carle. Copyright © 1984 by The Eric Carle Corporation.

Printed in the United States of America

ISBN 0-15-352222-4

11 12 13 14 15 1678 15 14 13 12 11 10

4500266687

Senior Author

Evan M. Maletsky
Professor of Mathematics
Montclair State University
Upper Montclair, New Jersey

Authors

Angela Giglio Andrews
Math Teacher, Scott School
Naperville District #203
Naperville, Illinois

Jennie M. Bennett
Houston Independent School District
Houston, Texas

Grace M. Burton
Professor, Watson School of Education
University of North Carolina at Wilmington
Wilmington, North Carolina

Lynda A. Luckie
K–12 Mathematics Coordinator
Gwinnett County Public Schools
Lawrenceville, Georgia

Joyce C. McLeod
Visiting Professor
Rollins College
Winter Park, Florida

Vicki Newman
Classroom Teacher
McGaugh Elementary School
Los Alamitos Unified School District
Seal Beach, California

Tom Roby
Associate Professor of Mathematics
California State University
Hayward, California

Janet K. Scheer
Executive Director
Create A Vision
Foster City, California

Program Consultants and Specialists

Janet S. Abbott
Mathematics Consultant
California

Elsie Babcock
Director, Mathematics and Science Center
Mathematics Consultant
Wayne Regional Educational Service Agency
Wayne, Michigan

William J. Driscoll
Professor of Mathematics
Department of Mathematical Sciences
Central Connecticut State University
New Britain, Connecticut

Lois Harrison-Jones
Education and Management Consultant
Dallas, Texas

Rebecca Valbuena
Language Development Specialist
Stanton Elementary School
Glendora, California

Unit 1

ADDITION AND SUBTRACTION CONCEPTS

Why Learn Math? .... A–D
Getting Ready: Patterns, Numbers, and Graphs E–1

School-Home Connection 1A–1B

CHAPTER 1 — Addition Concepts 1
Check What You Know 2
1 Model Addition Stories ✋HANDS ON 3
2 Use Symbols to Add 5
3 **Algebra:** Add 0 7
4 Problem Solving Strategy: Write a Number Sentence 9
Extra Practice 11
Review/Test • Standardized Test Prep 12–13
Math Game: Pet Numbers 14

Theme: My Favorite Animals

CHAPTER 2 — Using Addition 15
Check What You Know 16
1 **Algebra:** Add in Any Order ✋HANDS ON 17
2 Ways to Make 7 and 8 ✋HANDS ON 19
3 Ways to Make 9 and 10 ✋HANDS ON 21
4 Vertical Addition 23
5 Problem Solving Strategy: Make a Model 25
Extra Practice 27
Review/Test • Standardized Test Prep 28–29
Math Game: Tic-Tac-Sum 30

Theme: Getting to Know Us

CHAPTER 3

Subtraction Concepts 31
Check What You Know 32
1 Model Subtraction Stories HANDS ON 33

Theme: 2 Use Symbols to Subtract 35
At the 3 **Algebra:** Write Subtraction Sentences 37
Beach 4 Problem Solving Strategy: Make a Model 39
5 **Algebra:** Subtract All or Zero 41
Extra Practice 43
Review/Test • Standardized Test Prep 44–45
Math Game: Numbers in the Sand 46

CHAPTER 4

Using Subtraction 47
Check What You Know 48
1 Take Apart 7 and 8 HANDS ON 49

Theme: 2 Take Apart 9 and 10 HANDS ON 51
In the 3 Vertical Subtraction 53
Classroom 4 Subtract to Compare 55
5 Problem Solving Strategy: Draw a Picture 57
Extra Practice 59
Review/Test • Standardized Test Prep 60–61
It's in the Bag: Cat Pocket-Dots 62

Unit Wrap Up

Math Storybook: My Cat Biff .... A–D
Problem Solving On Location: Alabama 63
Challenge: Equals 64
Study Guide and Review 65
Performance Assessment 67
Technology: The Learning Site • Seashell Search 68

Unit 2

ADDITION AND SUBTRACTION FACTS TO 10

School-Home Connection 69A–69B

CHAPTER 5
Addition Strategies 69
Check What You Know 70
1 Count On 1 and 2 HANDS ON 71

Theme: Sea Life
2 Use a Number Line to Count On 73
3 Use Doubles HANDS ON 75
4 Problem Solving Strategy: Draw a Picture 77
Extra Practice 79
Review/Test • Standardized Test Prep 80–81
Math Game: Doubles Bubbles 82

CHAPTER 6
Addition Facts Practice 83
Check What You Know 84
1 Use the Strategies 85

Theme: On the Playground
2 Sums to 8 87
3 Sums to 10 89
4 **Algebra:** Follow the Rule 91
5 Problem Solving Strategy: Write a Number Sentence 93
Extra Practice 95
Review/Test • Standardized Test Prep 96–97
Math Game: Building Numbers 98

CHAPTER 7

Theme: Things in the Air

Subtraction Strategies	99
Check What You Know	100
1 Use a Number Line to Count Back 1 and 2	101
2 Use a Number Line to Count Back 3	103
3 **Algebra:** Relate Addition and Subtraction 🖐 HANDS ON	105
4 Problem Solving Strategy: Draw a Picture	107
Extra Practice	109
Review/Test • Standardized Test Prep	110–111
Math Game: Up, Up, and Away	112

CHAPTER 8

Theme: Fun Food

Subtraction Facts Practice	113
Check What You Know	114
1 Use the Strategies	115
2 Subtraction to 10	117
3 **Algebra:** Follow the Rule	119
4 Fact Families to 10 🖐 HANDS ON	121
5 Problem Solving Skill: Choose the Operation	123
Extra Practice	125
Review/Test • Standardized Test Prep	126–127
It's in the Bag: Math Under the Sea	128

Unit Wrap Up

Math Storybook: Under the Sea	A–F
Problem Solving On Location: Tennessee	129
Challenge: Missing Parts	130
Study Guide and Review	131
Performance Assessment	133
Technology: Calculator • Add and Subtract	134

Unit 4

GEOMETRY AND ADDITION AND SUBTRACTION TO 20

School-Home Connection 249A–249B

CHAPTER 15
Solid Figures and Plane Shapes **249**

Theme: Shapes in Our World

Check What You Know 250
1 Solid Figures 👋 HANDS ON 251
2 Faces and Vertices 👋 HANDS ON 253
3 Plane Shapes on Solid Figures 👋 HANDS ON ... 255
4 Sort and Identify Plane Shapes 👋 HANDS ON .. 257
5 Problem Solving Strategy: Make a Model 259
Extra Practice 261
Review/Test • Standardized Test Prep 262–263
Math Game: Make That Shape 264

CHAPTER 16
Spatial Sense **265**

Theme: On the Move

Check What You Know 266
1 Open and Closed 267
2 Problem Solving Skill: Use a Picture 269
3 Give and Follow Directions 271
4 Symmetry 👋 HANDS ON 273
5 Slides and Turns 👋 HANDS ON 275
Extra Practice 277
Review/Test • Standardized Test Prep 278–279
Math Game: On the Map 280

CHAPTER 17
Patterns **281**

Theme: Patterns All Around Us

Check What You Know 282
1 **Algebra:** Describe and Extend Patterns ... 283
2 **Algebra:** Pattern Units 👋 HANDS ON 285
3 **Algebra:** Make New Patterns 👋 HANDS ON .. 287
4 Problem Solving Skill: Correct a Pattern 289
5 Problem Solving Skill: Transfer Patterns 291
Extra Practice 293
Review/Test • Standardized Test Prep 294–295
Math Game: Pattern Play 296

CHAPTER 18
Addition Facts and Strategies 297
Check What You Know 298

Theme: Wildlife
1 Doubles and Doubles Plus 1 ✋HANDS ON 299
2 10 and More ✋HANDS ON 301
3 Make 10 to Add ✋HANDS ON 303
4 Use Make a 10 ✋HANDS ON 305
5 **Algebra:** Add 3 Numbers 307
6 Problem Solving Skill: Use Data from a Table 309
Extra Practice 311
Review/Test • Standardized Test Prep 312–313
Math Game: Ten Plus 314

CHAPTER 19
Subtraction Facts and Strategies 315
Check What You Know 316

Theme: In the Garden
1 Use a Number Line to Count Back 317
2 Doubles Fact Families 319
3 **Algebra:** Related Addition and Subtraction Facts 321
4 Problem Solving Skill: Estimate Reasonable Answers 323
Extra Practice 325
Review/Test • Standardized Test Prep 326–327
Math Game: Fact Family Bingo 328

CHAPTER 20
Addition and Subtraction Practice 329
Check What You Know 330

Theme: Arctic Life
1 Practice the Facts 331
2 Fact Families to 20 333
3 **Algebra:** Ways to Make Numbers to 20 ✋HANDS ON 335
4 Problem Solving Strategy: Make a Model 337
Extra Practice 339
Review/Test • Standardized Test Prep 340–341
It's in the Bag: The Hungry Prince's Crown 342

Unit Wrap Up
Math Storybook: The Hungry Prince .... A–H
Problem Solving On Location: New York 343
Challenge: Repeated Addition 344
Study Guide and Review 345
Performance Assessment 347
Technology: The Learning Site • Addition Surprise 348

xi

Unit 5

MONEY, TIME, AND FRACTIONS

School-Home Connection 349A–349B

CHAPTER 21 — Fractions 349
Theme: At the Pizza Party

Check What You Know 350
1 Halves 351
2 Fourths 353
3 Thirds 355
4 Problem Solving Strategy: Use Logical Reasoning 357
5 Parts of Groups 359
Extra Practice 361
Review/Test • Standardized Test Prep 362–363
Math Game: Pizza Party 364

CHAPTER 22 — Counting Pennies, Nickels, and Dimes 365
Theme: It's in the Bank

Check What You Know 366
1 Pennies and Nickels 👋 HANDS ON 367
2 Pennies and Dimes 👋 HANDS ON 369
3 Count Groups of Coins 371
4 Count Collections 373
5 Problem Solving Strategy: Make a List 375
Extra Practice 377
Review/Test • Standardized Test Prep 378–379
Math Game: Finding Coins 380

CHAPTER 23 — Using Money 381
Theme: What's for Sale?

Check What You Know 382
1 Trade Pennies, Nickels, and Dimes 👋 HANDS ON 383
2 Quarters 👋 HANDS ON 385
3 Half Dollar and Dollar 👋 HANDS ON 387
4 Compare Values 389
5 Same Amounts 👋 HANDS ON 391
6 Problem Solving Strategy: Act It Out 393
Extra Practice 395
Review/Test • Standardized Test Prep 396–397
Math Game: Shopping Basket 398

xii

CHAPTER 24 — Telling Time 399

Theme: What Time Is It?

- Check What You Know 400
- 1 Read a Clock HANDS ON 401
- 2 Problem Solving Skill: Use Estimation 403
- 3 Time to the Hour HANDS ON 405
- 4 Tell Time to the Half Hour HANDS ON 407
- 5 Practice Time to the Hour and Half Hour 409
- Extra Practice 411
- Review/Test • Standardized Test Prep 412–413
- Math Game: Clock Switch 414

CHAPTER 25 — Time and Calendar 415

Theme: All in My Day

- Check What You Know 416
- 1 Use a Calendar 417
- 2 Daily Events 419
- 3 Problem Solving Strategy: Make a Graph 421
- 4 Read a Schedule 423
- 5 Problem Solving Skill: Make Reasonable Estimates 425
- Extra Practice 427
- Review/Test • Standardized Test Prep 428–429
- It's in the Bag: Brown-Bag Grandfather Clock 430

Unit Wrap Up

- **Math Storybook:** Is It Time? .... A–H
- **Problem Solving On Location:** Louisiana 431
- **Challenge:** Writing Fractions 432
- **Study Guide and Review** 433
- **Performance Assessment** 435
- **Technology:** The Learning Site • Willy the Watchdog 436

Unit 6

MEASUREMENT, OPERATIONS, AND DATA

School-Home Connection 437A–437B

CHAPTER 26 — Length 437

Theme: Arts and Crafts

Check What You Know 438
1 Compare Lengths HANDS ON 439
2 Use Nonstandard Units HANDS ON 441
3 Inches HANDS ON 443
4 Inches and Feet HANDS ON 445
5 Centimeters HANDS ON 447
6 Problem Solving Skill: Make Reasonable Estimates 449
Extra Practice 451
Review/Test • Standardized Test Prep 452–453
Math Game: Ruler Race 454

CHAPTER 27 — Weight 455

Theme: At the Farm

Check What You Know 456
1 Use a Balance HANDS ON 457
2 Pounds HANDS ON 459
3 Kilograms HANDS ON 461
4 Problem Solving Strategy: Predict and Test 463
Extra Practice 465
Review/Test • Standardized Test Prep 466–467
Math Game: Gram Grab 468

CHAPTER 28 — Capacity 469

Theme: Fill It Up

Check What You Know 470
1 Nonstandard Units HANDS ON 471
2 Cups, Pints, and Quarts HANDS ON 473
3 Liters HANDS ON 475
4 Temperature 477
5 Problem Solving Skill: Choose the Measuring Tool 479
Extra Practice 481
Review/Test • Standardized Test Prep 482–483
Math Game: How Many Cups? 484

CHAPTER 29 — Adding and Subtracting 2-Digit Numbers 485

Theme: In the Rain Forest

Check What You Know 486
1 Use Mental Math to Add Tens 487
2 Add Tens and Ones HANDS ON 489
3 Add Money 491
4 Use Mental Math to Subtract Tens 493
5 Subtract Tens and Ones HANDS ON 495
6 Subtract Money 497
7 Problem Solving Skill: Make Reasonable Estimates 499
Extra Practice 501
Review/Test • Standardized Test Prep 502–503
Math Game: Math Path 504

CHAPTER 30 — Probability 505

Theme: Chances

Check What You Know 506
1 Certain or Impossible 507
2 More Likely, Less Likely 509
3 Equally Likely 511
4 Problem Solving Skill: Make a Prediction 513
Extra Practice 515
Review/Test • Standardized Test Prep 516–517
It's in the Bag: Cool Cat Hat 518

Unit Wrap Up

Math Storybook: Cat's Cool Hat ..... A-H
Problem Solving On Location: West Virginia 519
Challenge: Area 520
Study Guide and Review 521
Performance Assessment 523
Technology: Calculator • Find the Greatest Sum 524
Picture Glossary 525

Why Learn Math?

Look at each picture.
Decide what math skill is being used.

Explain It What are some things you do to use math?

Practice What You Learn

IT'S IN THE BAG

PROJECT You will make a math facts vest.

You Will Need

- Large brown bag
- Pattern tracer
- Scissors
- Construction paper
- Glue
- Tape
- Crayons

Directions

1. Pop out the sides of the bag. Lay the bag flat.
2. Trace the vest pattern 2 times, once on each side of the bag.
3. Trace two arm holes. Cut out two arm holes from two layers of the bag.
4. Cut the vest front from only the top layer of the bag.
5. Tape the front of the bag to the back of the bag. Make pockets. Use construction paper.

Show What You Learn

The checklist shows what you will do when you take a test.

I will:

- listen carefully.
- read carefully.
- follow directions.
- mark answers carefully.
- begin where told.
- begin when told.
- pay attention only to the teacher and the test.
- do the best I can.

_____

Name

Name _____

1. Color the triangles red.
   Color the squares and rectangles blue.
   Color the circles yellow.

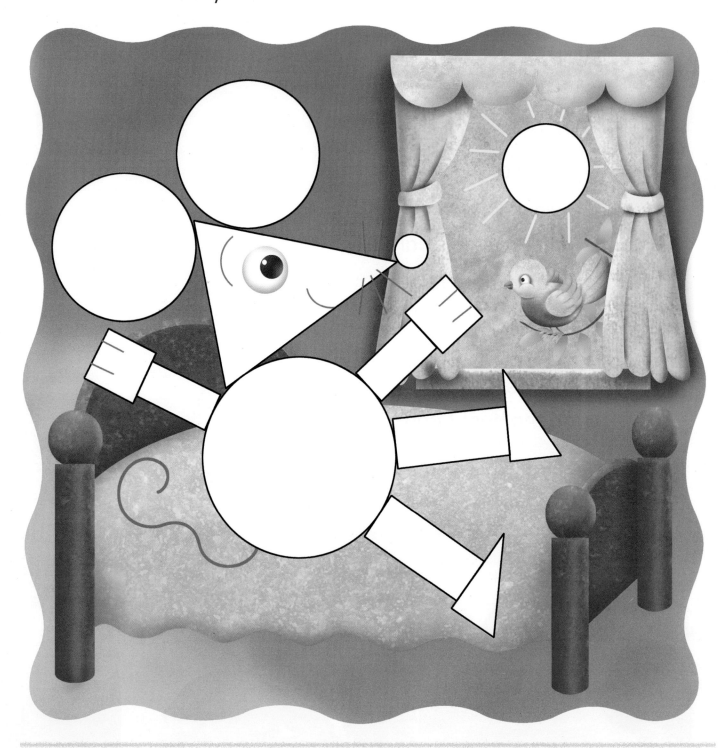

2. Draw what comes next in this pattern.

REVIEW: Shapes and Patterns

Name _____

1. Draw 1 sock for each shoe.

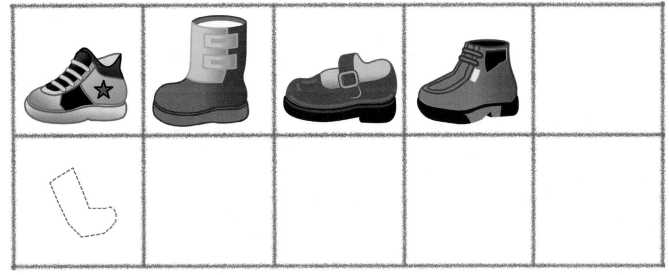

Count the objects. Write the numbers.

2. 	3. 	4. 	5.	6.
5	___	___	___	___
five	two	eight	three	six
7. 	8. 	9. 	10. 	11.
___	___	___	___	___
seven	nine	ten	four	one

REVIEW: One to One Correspondence/Numbers to 10

F

Name _____

1. Circle the row that has more.

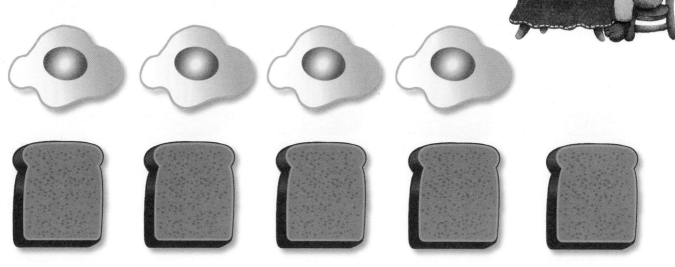

2. Circle the row that has fewer.

Write the missing number.

3.

3   4   ___

4.

___   9   10

G

REVIEW: More, Fewer/Order Through 10

Name _____

1. Get some ▪ and ▫. Sort.
   Make a graph.
   Write how many.

   How many ▪ and ▫?

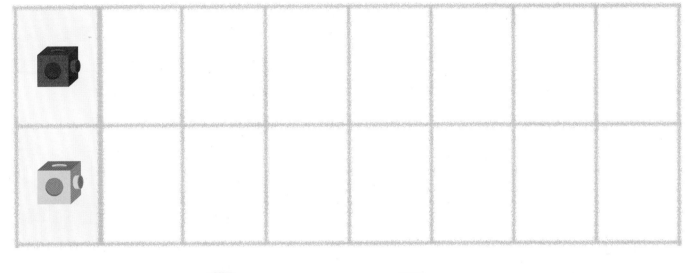

▪ _____   ▫ _____

2. Write how many.

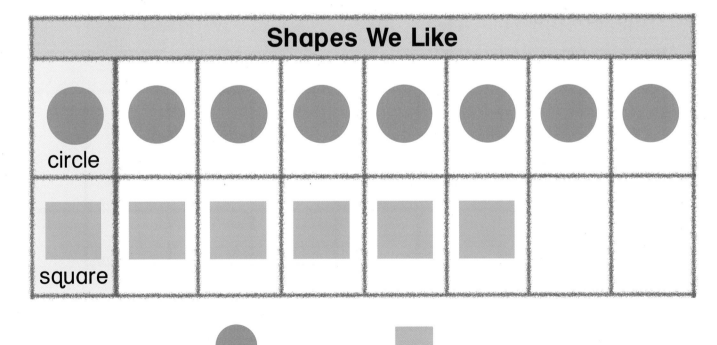

● _____   ■ _____

3. Which shape do more children like?
   Circle that shape.

REVIEW: Concrete Graphs/Picture Graphs

H

Name _____

1. Write how many.

What We Like to Do	
📖	|||||
🧊	||||

2. Which do more children like to do? Circle that picture.

3. Write how many.

Boys and Girls Absent							
👦							
👧							

REVIEW: Tally Table/Bar Graph

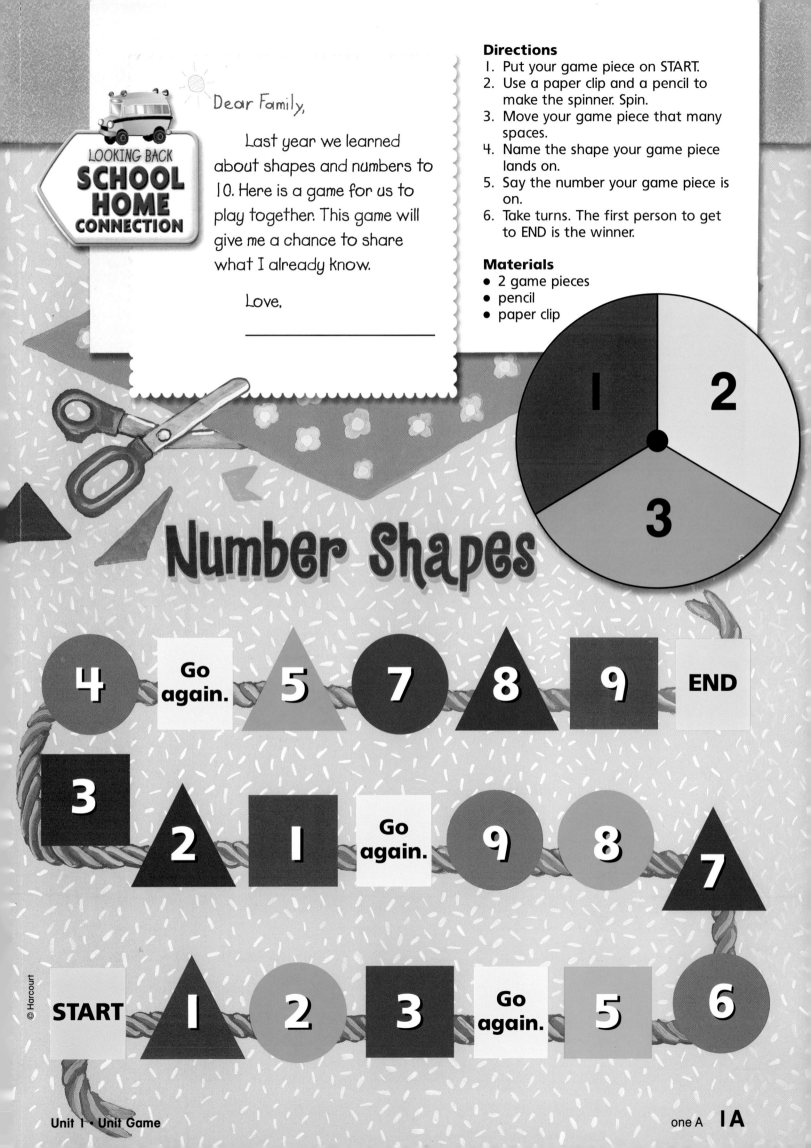

Dear Family,

During the next few weeks, we will learn to add and subtract through 10. Here is important math vocabulary and a list of books to share.

Love,

_____

Vocabulary
addition sentence
sum
subtraction sentence
difference

Vocabulary Power

$4 + 1 = 5$

↑ The sum tells how many in all.

$4 + 1 = 5$ is an addition sentence.

$3 - 1 = 2$

↑ The difference tells how many are left.

$3 - 1 = 2$ is a subtraction sentence.

BOOKS TO SHARE

To read about addition and subtraction with your child, look for these books in your library.

Anno's Counting Book, by Mitsumasa Anno, HarperCollins, 1992.

Fish Eyes: A Book You Can Count On, by Lois Ehlert, Harcourt, 2001.

Roll Over! A Counting Song, illustrated by Merle Peek, Clarion, 1999.

Ten Little Mice, by Joyce Dunbar, Gulliver, 1992.

 Visit *The Learning Site* for additional ideas and activities. www.harcourtschool.com

CHAPTER 1 Addition Concepts

FUN FACTS

Cats have 4 sets of whiskers. Look on the chin, cheek, wrist, and eyebrow.

Theme: My Favorite Animals

Name _____

✓ Check What You Know

Zero

Count the . Write how many .

1.  __2__

2.  __3__

Count and Write the Numbers to 10

Count the ●. Write the number.

3.  __9__

4.  __10__

Model Addition

Use 🟦 to show how many in all.

Draw the 🟦.
Write the number that tells how many in all.

5.

        4   1   __5__

6.

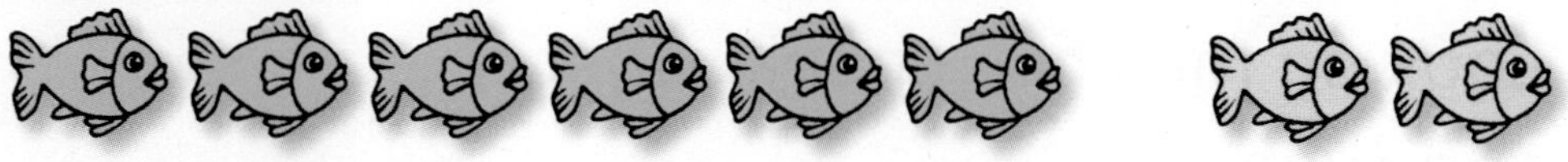

        6   2   __8__

Name _____

Model Addition Stories

Explore (Hands On)

Use ◯ to show the story.
Draw the ◯. Write how many in all.

"There are 2 birds in all."

1 big bird   1 little bird   __2__ in all

1.
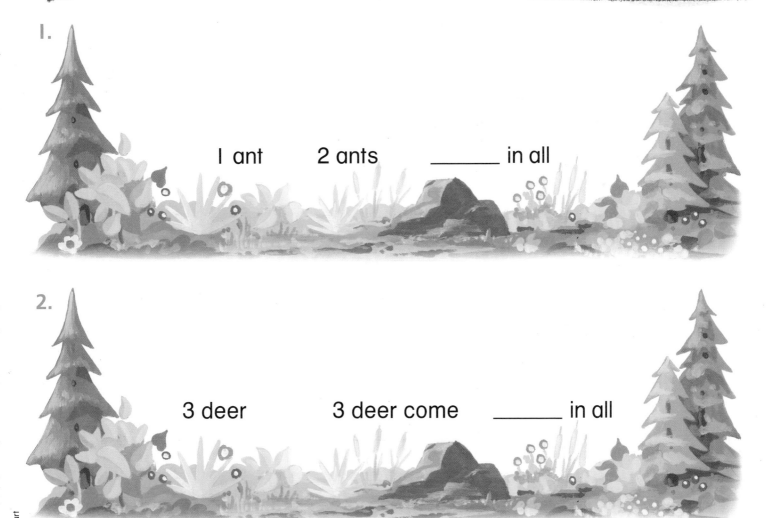

1 ant   2 ants   _____ in all

2.

3 deer   3 deer come   _____ in all

Explain It • Daily Reasoning

What happens when more is added to a group?
Explain how you know.

Chapter 1 • Addition Concepts

Practice

Use ◯ to show the story.
Draw the ◯. Write how many in all.

1.
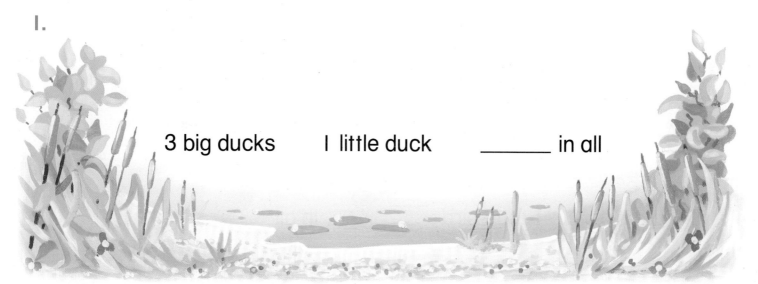
3 big ducks   1 little duck   _____ in all

2.
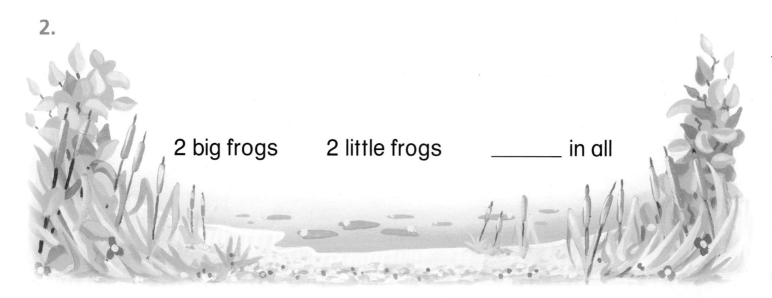
2 big frogs   2 little frogs   _____ in all

3.

1 bee   4 bees come   _____ in all

HOME ACTIVITY • Have your child tell an addition story for each picture.

Name: Hannah

Use Symbols to Add

Vocabulary
plus +
equals =
sum

Learn

3 + 2 = 5
↑ ↑ ↑
plus equals sum

Check

Add. Write the sum.

1.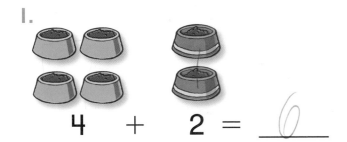

   4 + 2 = 6

2. 3 + 3 = 6

3.

   2 + 2 = 4

4.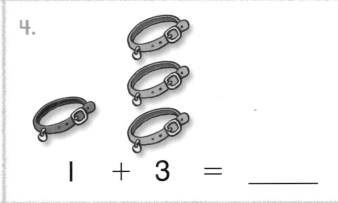

   1 + 3 = ___

5.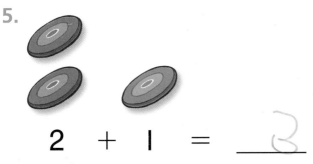

   2 + 1 = 3

6.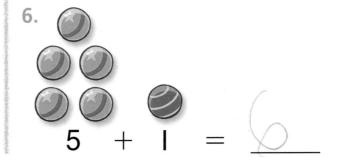

   5 + 1 = 6

Explain It • Daily Reasoning

How could you find the sum for 1 + 4 without using pictures?

Chapter 1 • Addition Concepts

Practice and Problem Solving

Add. Write the sum.

1.

   3 + 1 = 4

2.

   2 + 3 = 5

3.

   4 + 1 = 5

4.

   2 + 4 = 6

Problem Solving

Application

Use counters to show the story. Write the sum.

5. There are 5 black kittens. There is 1 gray kitten. How many kittens are there in all? _____ kittens

 Write About It • Draw a picture. Show an addition story about a large dog and some small dogs. Write the sum.

HOME ACTIVITY • Ask your child to tell how he or she found each sum. Then have him or her draw pictures to show 1 + 5 and tell the sum. (6)

Name _____

Algebra: Add 0

Vocabulary
zero 0

Learn

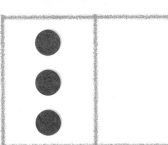

Any number plus 0 equals the same number.

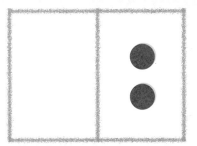

3 + 0 = __3__

0 + 2 = __2__

Check

Write the sum.

1.
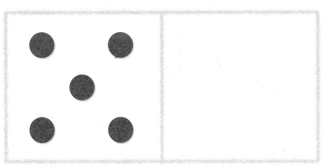

5 + 0 = ____

2.

0 + 1 = ____

3.
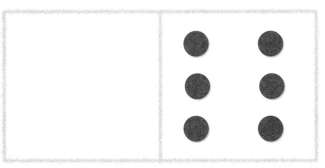

0 + 6 = ____

4.
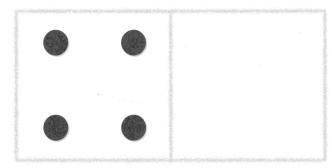

4 + 0 = ____

Explain It • Daily Reasoning

What happens when 0 is added to a number? Why?

Name ______________________________

Algebra: Add in Any Order

Vocabulary
sum

HANDS ON

Explore

2 + 1 = 3
sum

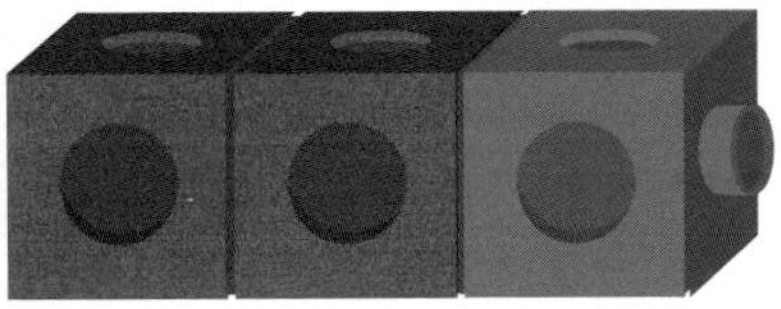

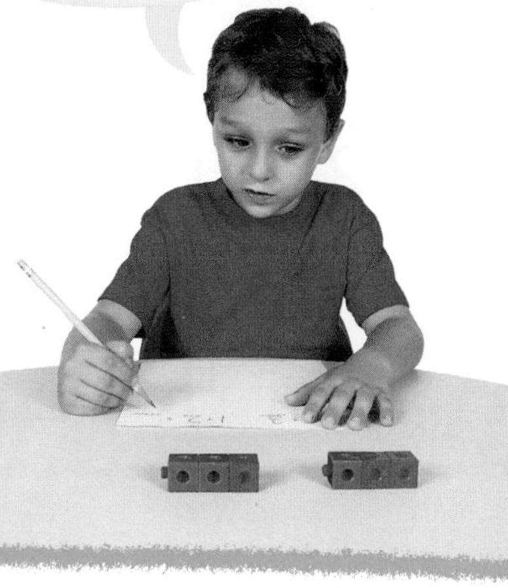
You can add in any order and get the same sum.

1 + 2 = 3
sum

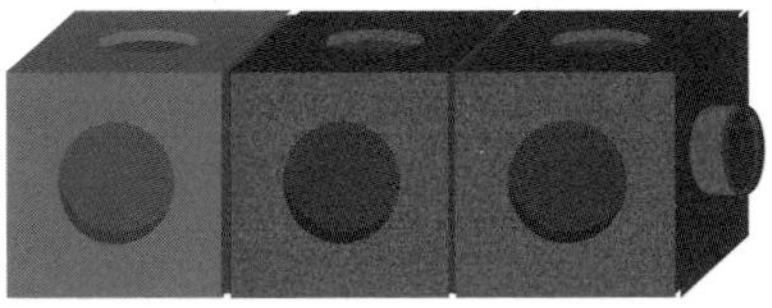

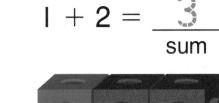

Connect

Use and to add. Write each sum.
Color to match.

1. 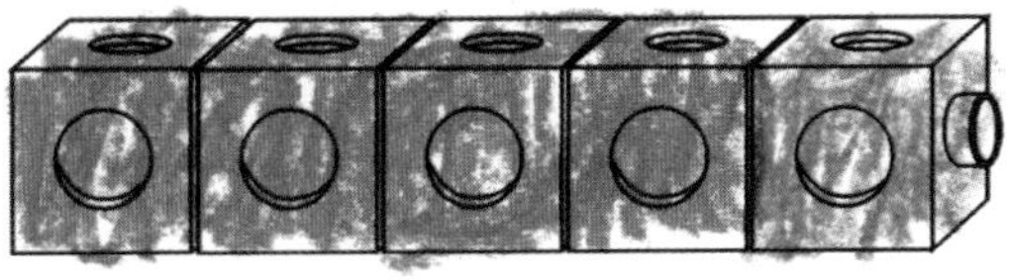

4 + 1 = ____

2. 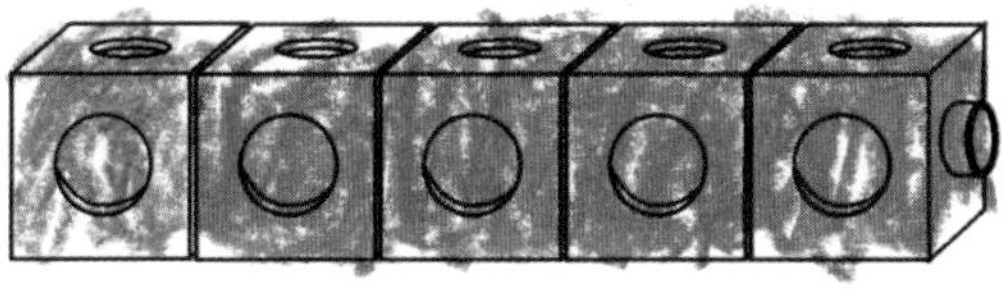

1 + 4 = ____

3. 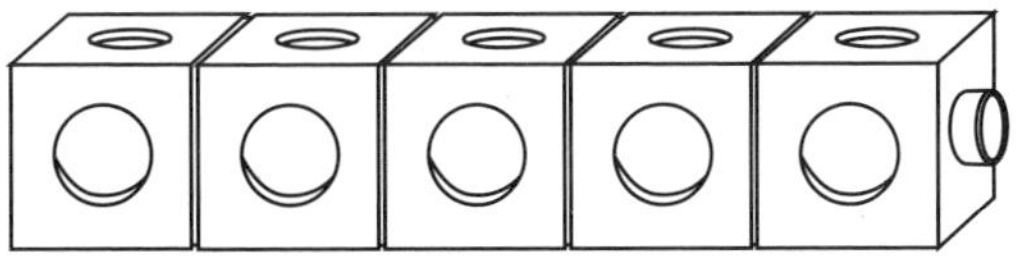

3 + 2 = ____

4. 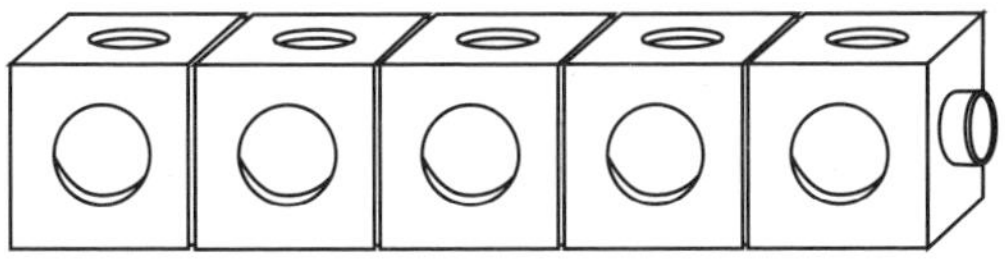

2 + 3 = ____

5. 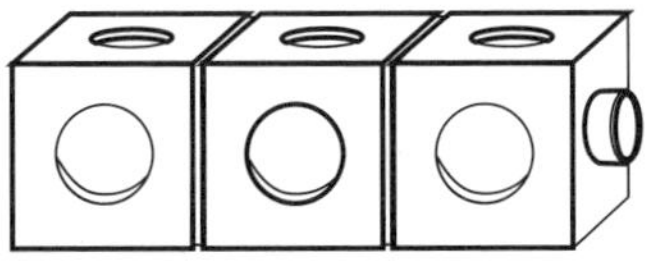

3 + 0 = ____

6. 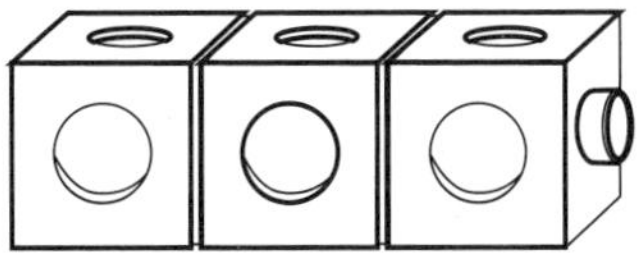

0 + 3 = ____

Explain It • Daily Reasoning

What happens to the sum when you change the order of the numbers you are adding? Use and to prove your answer.

Practice and Problem Solving

Use 🎲 and 🎲 to add. Circle the addition sentences in each row that have the same sum.

1. (3 + 1 = __4__)   2. 2 + 1 = __3__   3. (1 + 3 = __4__)

4. 4 + 2 = ____   5. 2 + 4 = ____   6. 1 + 4 = ____

7. 3 + 3 = ____   8. 4 + 0 = ____   9. 0 + 4 = ____

10. 4 + 1 = ____   11. 5 + 1 = ____   12. 1 + 5 = ____

13. 2 + 3 = ____   14. 2 + 2 = ____   15. 3 + 2 = ____

Problem Solving

Application

Circle your answer.

16. Bob has 1 🎲 and 4 🎲.
    Pat has 4 🎲 and 1 🎲.
    Do they have the same number of cubes?

    Yes   No

    Prove your answer.

 Write About It • Show two ways to add 1 and 5. Tell why the sums are the same.

⬠ **HOME ACTIVITY** • Have your child use small objects to show 3 + 1 and 1 + 3 and then tell you why the two sums are the same.

Name _____

Ways to Make 7 and 8

Vocabulary
plus
equals

Explore

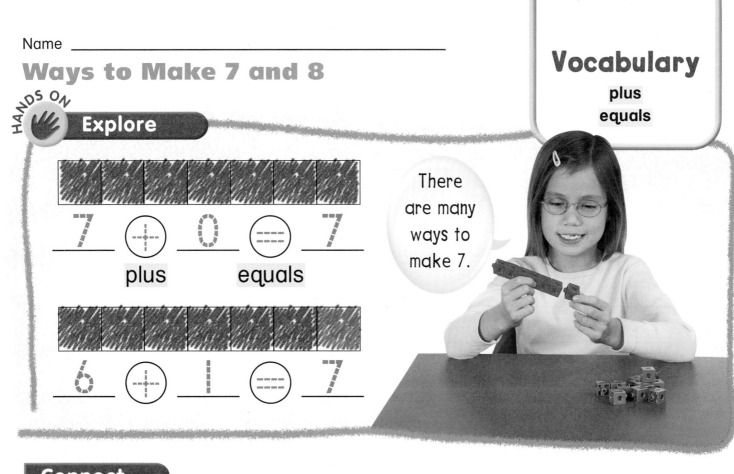

There are many ways to make 7.

7 + 0 = 7
plus   equals

6 + 1 = 7

Connect

Use 🟥 and 🟦 to make 7.
Color. Write the addition sentence.

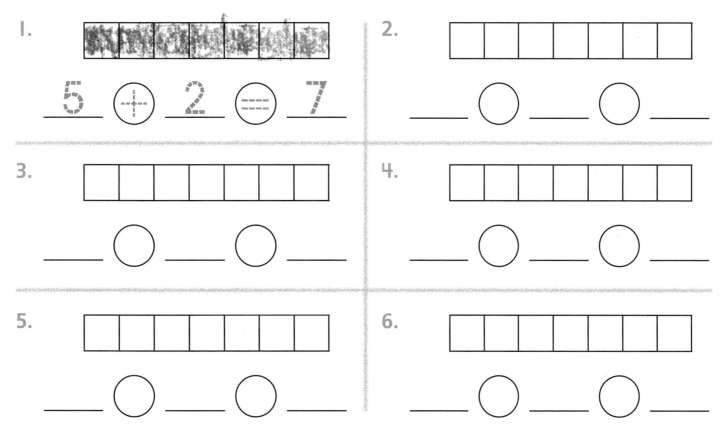

1. 5 + 2 = 7

2. ___ ○ ___ ○ ___

3. ___ ○ ___ ○ ___

4. ___ ○ ___ ○ ___

5. ___ ○ ___ ○ ___

6. ___ ○ ___ ○ ___

Explain It • Daily Reasoning

If you have 3 🟥, how many 🟦 do you need to make 7? Use 🟥 and 🟦 to prove your answer.

Chapter 2 • Using Addition

Practice and Problem Solving

Use ◼ and ◼ to make 8.
Color. Write the addition sentence.

This is one way to make 8.

1. [colored row of 8 squares]

   _7_ ⊕ _1_ ⊖ _8_

2.

   ___ ◯ ___ ◯ ___

3.

   ___ ◯ ___ ◯ ___

4.

   ___ ◯ ___ ◯ ___

5.

   ___ ◯ ___ ◯ ___

6.

   ___ ◯ ___ ◯ ___

7.

   ___ ◯ ___ ◯ ___

8.

   ___ ◯ ___ ◯ ___

Problem Solving

Visual Thinking

9. Write an addition sentence that tells about the picture.

   ___ ◯ ___ ◯ ___

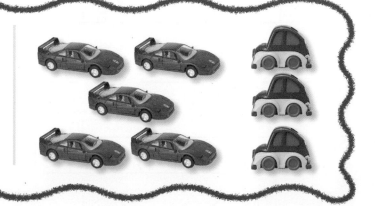

Write About It • How can you show 8 a different way? Draw cars to show how. Write the addition sentence.

⬟ **HOME ACTIVITY** • Have your child use small objects to show different ways to make 7 and 8.

Name _____

Ways to Make 9 and 10

Explore

This is one way to make 9.

Connect

Use Workmat 7, ●, and ○ to make 9.
Draw and color. Write the addition sentence.

1.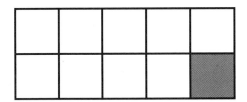

2.

3.

4.

5.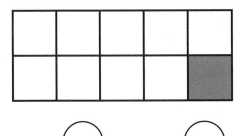

6.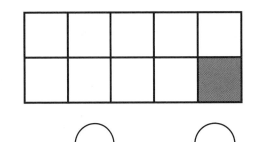

Explain It • Daily Reasoning

How could you use ● and ○ to make 10?

Chapter 2 • Using Addition

twenty-one **21**

Practice and Problem Solving

Use Workmat 7, ●, and ○ to make 10.
Draw and color. Write the addition sentence.

1.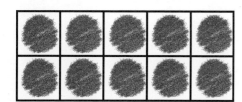

   _10_ ⊕ _0_ ⊜ _10_

2.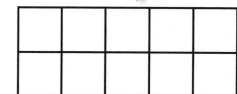

   ___ ○ ___ ○ ___

3.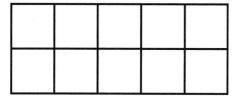

   ___ ○ ___ ○ ___

4.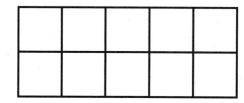

   ___ ○ ___ ○ ___

5.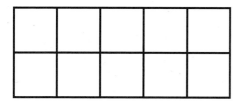

   ___ ○ ___ ○ ___

6.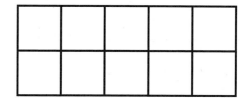

   ___ ○ ___ ○ ___

Problem Solving
Application

7. You have 5 pennies. How many more pennies do you need to make 10?

   _____ more pennies

Write About It • You have 7 pennies. You need 10 pennies. Draw to show how many more pennies you need. Write the addition sentence.

 HOME ACTIVITY • Have your child use small objects to show different combinations that make 9 and 10.

22 twenty-two

Name _____

Vertical Addition

Learn

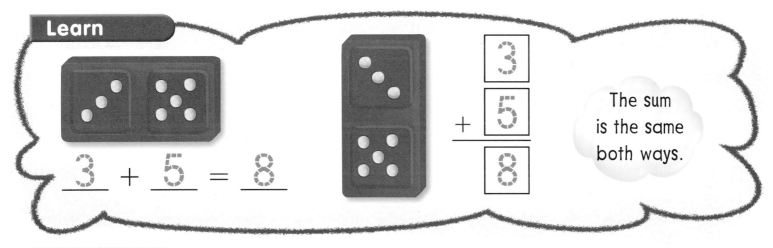

____3____ + ____5____ = ____8____

The sum is the same both ways.

Check

Write the numbers to match the dots.
Write the sum.

1.

   ____ + ____ = ____

2.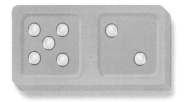

   ____ + ____ = ____

3.

   ____ + ____ = ____

Explain It • Daily Reasoning

How are the problems in each row alike?
How are they different? Explain.

Chapter 2 • Using Addition

Practice and Problem Solving

Write the numbers to match the dots.
Write the sum.

1. 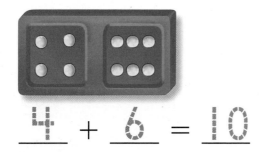 $\underline{4} + \underline{6} = \underline{10}$

Write the sum.

2. 4
   +2

3. 6
   +2

4. 5
   +5

5. 7
   +2

6. 0
   +5

7. 2
   +2

8. 1
   +2

9. 5
   +4

10. 2
    +5

11. 4
    +6

12. 8
    +1

13. 0
    +6

---

14. 3
    +2

15. 2
    +3

16. 8
    +0

17. 3
    +6

18. 7
    +3

19. 4
    +0

Problem Solving
Application

20. Each person has 5 red pencils. They get some blue pencils. How many pencils does each person have now?

Pat	Jack	Al
5	5	5
+0	+1	+2
☐	☐	☐

Write About It • Look at Exercise 20. What pattern do you see?

🏠 HOME ACTIVITY • Write addition problems both across and down for your child to solve.

Name _____

Problem Solving Strategy
Make a Model

How much do these cost altogether?

UNDERSTAND

What do you know?

Marbles cost __5__ ¢.   A giraffe costs __1__ ¢.

PLAN

How do you solve this problem?

Make a model.

SOLVE

Show 5 pennies.
Then show 1 penny.
Count the pennies.

__6__ ¢

CHECK

Explain why you think your answer is right.

Use 🪙 to show each price.

Draw the 🪙. Write how many there are in all.

THINK: What do I know?

1. How much do you spend for both?

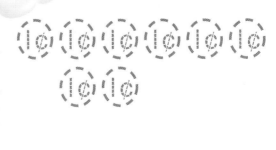

_____ ¢

Chapter 2 • Using Addition

twenty-five **25**

Name _____

MATH GAME

Tic-Tac-Sum

Play with a partner.

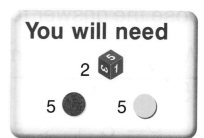
You will need

1. One player uses ●.
   The other player uses ○.
2. Toss two 🎲.
3. Find the sum.
   Cover that number with a counter.
4. Your turn is over if that number is already covered.
5. The first player to get 4 counters in a row wins.

CHAPTER 3

Subtraction Concepts

FUN FACTS

Leatherback turtles are as long as 8 of your math books.

Theme: At the Beach

Practice

Use ● to show the story. Draw the ●.
Cross out how many go away. Write how many are left.

1. 4 bees   4 fly away   _____ are left

2. 2 gulls   1 walks away   _____ is left

3. 5 children   2 walk away   _____ are left

HOME ACTIVITY • Have your child use objects to show the subtraction stories on this page.

Name _____

Use Symbols to Subtract

Vocabulary
minus –
equals =
difference

Learn

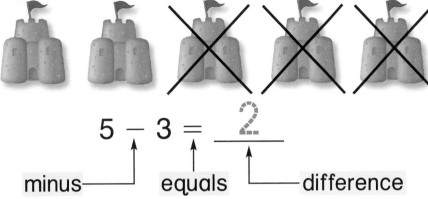

$5 - 3 = \underline{2}$

minus — equals — difference

Check

Cross out pictures to subtract.
Write the difference.

1.  $4 - 2 = \underline{}$

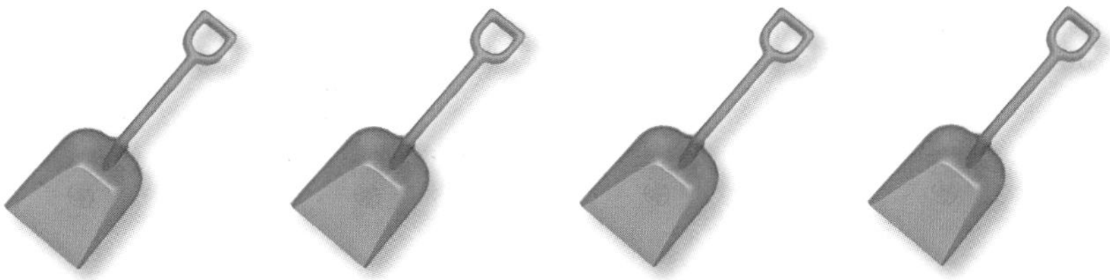

2.  $6 - 3 = \underline{}$

3.  $5 - 2 = \underline{}$

4.  $3 - 1 = \underline{}$

Explain It • Daily Reasoning

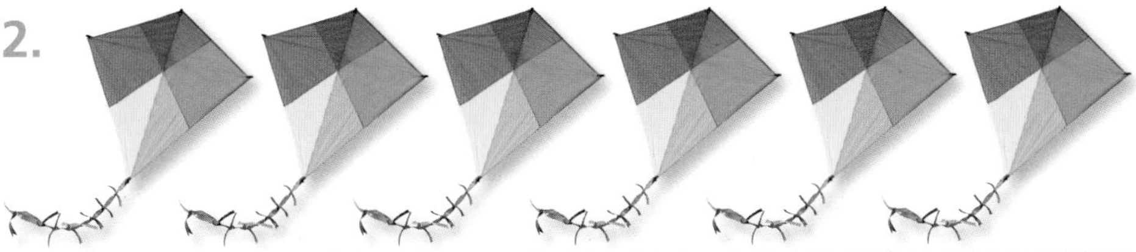

What does the minus sign mean?
What does the equal sign mean? Explain.

Chapter 3 • Subtraction Concepts

Practice and Problem Solving

Cross out pictures to subtract.
Write the difference.

1.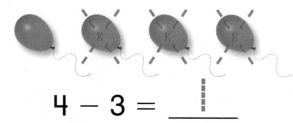

    4 − 3 = 1

2.

    6 − 2 = ____

3.

    3 − 2 = ____

4.

    4 − 1 = ____

5.

    6 − 5 = ____

6.

    6 − 4 = ____

7.

    5 − 4 = ____

8.

    2 − 1 = ____

Problem Solving
Logical Reasoning

Solve the riddle. Write the number.

9. I am greater than 4. I am less than 6. What number am I?

    ____

10. I am less than 3. I am greater than 1. What number am I?

    ____

 Write About It • Write a riddle about the number 4.

HOME ACTIVITY • Have your child draw pictures to show a subtraction problem. Then ask him or her to tell the difference.

Name _____

Algebra: Write Subtraction Sentences

Vocabulary
subtraction sentence

Learn

___6___ ◯−◯ ___2___ ◯=◯ ___4___

6 − 2 = 4 is a subtraction sentence.

Check

Write the subtraction sentence.

1.
___ ◯ ___ ◯ ___

2.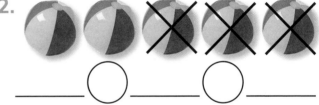
___ ◯ ___ ◯ ___

3.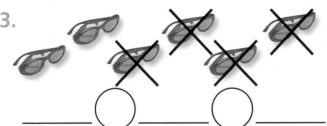
___ ◯ ___ ◯ ___

4.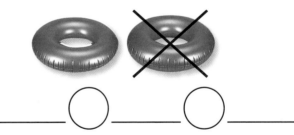
___ ◯ ___ ◯ ___

5.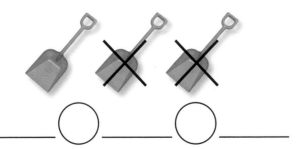
___ ◯ ___ ◯ ___

6.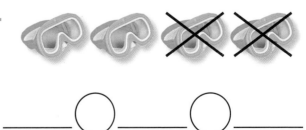
___ ◯ ___ ◯ ___

Explain It • Daily Reasoning

How can you use these numbers to write two different subtraction sentences? Explain.

1   3   4

Practice and Problem Solving

Write the subtraction sentence.

1.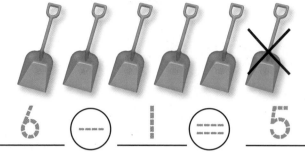
   6 − 1 = 5

2.
   ___ ◯ ___ ◯ ___

3.
   ___ ◯ ___ ◯ ___

4.
   ___ ◯ ___ ◯ ___

5.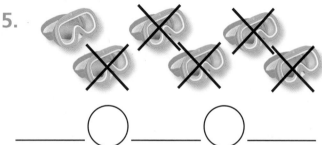
   ___ ◯ ___ ◯ ___

6.
   ___ ◯ ___ ◯ ___

Problem Solving
Visual Thinking

Circle the picture that shows the subtraction sentence.

7. 4 − 3 = 1

8. 3 − 1 = 2

Write About It • Look at Exercises 7 and 8. Explain why you circled the pictures you did.

HOME ACTIVITY • Have your child use objects to show subtraction stories. Then ask him or her to write the subtraction sentences.

38 thirty-eight

Name _____

Problem Solving Strategy
Make a Model

3 butterflies are in the tree.

1 flies away.

(How many are left?)

UNDERSTAND

What do you need to find out?
Circle the question.
What do you know?

There are __3__ butterflies.

__1__ flies away.

PLAN

How do you solve this problem?
Use counters. Draw 3 counters. Cross out 1.

SOLVE

There are __2__ butterflies left.

CHECK

Does your answer make sense?
Explain.

Use ● to subtract.
Draw the ●.
Write the difference.

1. Kathy finds 4 shells.
   She gives 2 away.
   How many shells
   does she have left?

   THINK: What do I need to find out?

   _____ shells

Chapter 3 • Subtraction Concepts

thirty-nine **39**

Problem Solving Practice

Use ● to subtract.
Draw the ●.
Write the difference.

THINK: How can I solve the problem?

Keep in Mind!
Understand
Plan
Solve
Check

1. Steve sees 6 pelicans.
   2 fly away.
   How many pelicans are left?

   _____ pelicans

2. Lisa sees 5 fish.
   3 swim away.
   How many fish are left?

   _____ fish

3. Tim finds 3 shells.
   He gives 1 to Joe.
   How many shells does he have left?

   _____ shells

4. Ann sees 2 sand castles.
   1 gets washed away.
   How many sand castles are left?

   _____ sand castle

HOME ACTIVITY • Make up story problems like the ones in this lesson. Have your child use objects or draw pictures to solve the problems.

Name _____

Algebra: Subtract All or Zero

Vocabulary
zero 0

Learn

6 − 0 = __6__

When you subtract zero, you have the same number left.

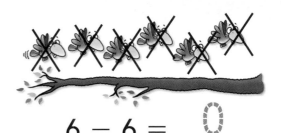

6 − 6 = __0__

When you subtract all, you have zero left.

Check

Write the difference.

1.

   3 − 0 = _____

2.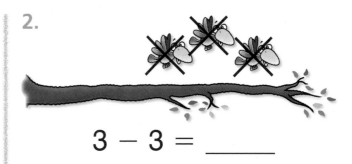

   3 − 3 = _____

3.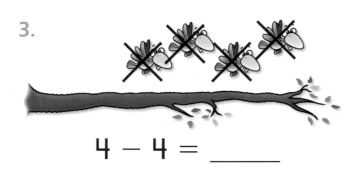

   4 − 4 = _____

4.

   4 − 0 = _____

5.

   2 − 0 = _____

6.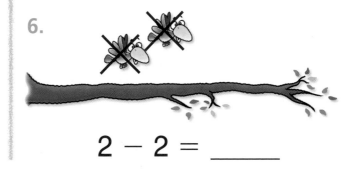

   2 − 2 = _____

Explain It • Daily Reasoning

What happens when you subtract all of a group? Why?
What happens when you subtract 0 from a group? Why?

Chapter 3 • Subtraction Concepts

forty-one **41**

Practice and Problem Solving

Write the difference.

1.
5 − 5 = 0

2.
6 − 0 = 6

3.
4 − 4 = ___

4.
2 − 0 = ___

5.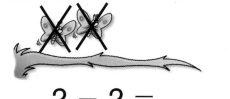
2 − 2 = ___

6.
1 − 1 = ___

7.
5 − 0 = ___

8.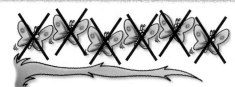
6 − 6 = ___

Problem Solving
Application

Write the subtraction sentence.

9. Maria sees 3 butterflies. All 3 fly away. How many are left?

___ ◯ ___ ◯ ___

 Write About It • Look at Exercise 9. Draw pictures to show your subtraction sentence. Explain if you subtracted all or zero.

HOME ACTIVITY • Have your child draw pictures to show 3 − 3 and 3 − 0. Ask your child to tell you how to find each difference.

42 forty-two

Name _____

Extra Practice

Cross out pictures to subtract.
Write the difference.

1.
4 – 3 = ____

2.
5 – 2 = ____

Write the subtraction sentence.

3.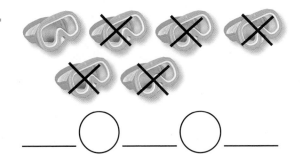
____ ◯ ____ ◯ ____

4.
____ ◯ ____ ◯ ____

Write the difference.

5.
4 – 0 = ____

6.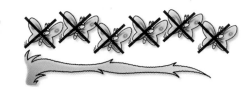
6 – 6 = ____

Problem Solving

Use ● to subtract.
Draw the ●.
Write the difference.

7. Joe sees 6 fish.
 2 swim away.
 How many fish are left?

____ fish

Chapter 3 • Subtraction Concepts

Name _____

✓ Review/Test

Concepts and Skills

Cross out pictures to subtract.
Write the difference.

1.

    6 − 3 = _____

2.

    5 − 4 = _____

Write the subtraction sentence.

3.

    ____ ◯ ____ ____

4.

    ____ ◯ ____ ____

Write the difference.

5.

    5 − 0 = _____

6.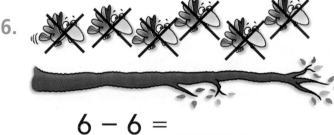

    6 − 6 = _____

Problem Solving

Use ● to subtract.
Draw the ●.
Write the difference.

7. 5 ants are on a log.
   3 ants crawl away.
   How many ants are left?

   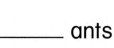 ants

44 forty-four

Name _____

⭐Standardized Test Prep
Chapters 1–3

Choose the answer for questions 1–4.

1. Which subtraction sentence tells about the picture?

   $6 - 6 = 0$ $6 - 3 = 3$ $6 - 4 = 2$ $6 - 1 = 5$
   ○   ○   ○   ○

2. What is the difference?

   $6 - 3 =$ _____

   3   6   8   11
   ○   ○   ○   ○

3. What is the difference for $4 - 0$?

   4   3   2   1
   ○   ○   ○   ○

4. There are 3 bees. 2 fly away. How many are left?

   0   1   3   5
   ○   ○   ○   ○

Show What You Know

5. Write an addition sentence to show the sum of 8. Use 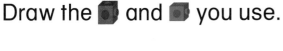 and  to explain your answer. Draw the 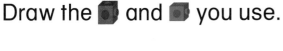 and  you use.

   _____ ○ _____ ○ _____

Chapter 3   forty-five **45**

MATH GAME

Numbers in the Sand

Play with a partner.

1. Put your ♟ at START.
2. Toss the 🎲.
3. Move your ♟ that many spaces.
4. Find the difference.
5. If you are not correct, lose a turn.
6. The first player to get to END wins.

You will need

2 ♟   🎲

CHAPTER 4
Using Subtraction

FUN FACTS

There are three primary colors: red, yellow, and blue.

Theme: In the Classroom

Name _____

✓ Check What You Know

Use Pictures to Subtract

Subtract. Write the numbers.

1.

   ___ – ___ = ___

2.

   ___ – ___ = ___

3.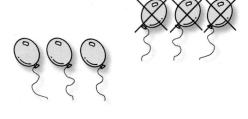

   ___ – ___ = ___

4.

   ___ – ___ = ___

Use Symbols to Subtract

Cross out pictures to subtract.
Write the difference.

5.

   $3 - 1 = $ ___

6.

   $6 - 1 = $ ___

7.

   $5 - 3 = $ ___

8.

   $4 - 3 = $ ___

Name _____

Take Apart 7 and 8

Vocabulary
subtraction sentence

Explore (Hands On)

I started with 7 cubes and then took one away.

7 − 1 = 6

subtraction sentence

Connect

Use ▬ to show all the ways to subtract from 7.
Complete the subtraction sentences.

1. 7 − __0__ = __7__

2. 7 − ___ = ___

3. 7 − ___ = ___

4. 7 − ___ = ___

5. 7 − ___ = ___

6. 7 − ___ = ___

7. 7 − ___ = ___

8. 7 − ___ = ___

Explain It • Daily Reasoning

Continue the pattern.
What comes next?
Explain how you know.

7−0=7
7−1=6
7−2=5

Chapter 4 • Using Subtraction

Practice and Problem Solving

Write the difference.

1.

    7 − 4 = __3__

2.

    7
    −4
    ———
    3

3. 10 − 0

4. 6 − 3

5. 4 − 2

6. 8 − 4

7. 9 − 2

8. 5 − 0

9. 3 − 2

10. 8 − 6

11. 9 − 0

12. 7 − 2

13. 5 − 4

14. 6 − 6

15. 10 − 5

16. 4 − 0

17. 7 − 7

18. 9 − 4

19. 8 − 0

20. 7 − 6

Problem Solving
Logical Reasoning

Solve. Write the number.

21. I am greater than 8.
    I am less than 10.
    What number am I?

22. I am less than 5.
    I am greater than 3.
    What number am I?

Write About It • Explain how you solve this riddle. I am greater than 0. I am less than 2. What number am I?

HOME ACTIVITY • Write subtraction sentences that go across, and have your child find the difference for each. Then ask your child to write the same problems going down.

Name _____

Subtract to Compare

Learn

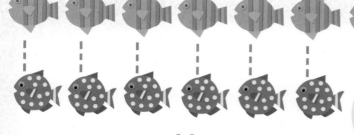

10 − 6 = __4__

There are more 🐟 than 🐟.
You can subtract to find how many more.

Check

Draw lines to match.
Subtract to find how many more.

1.

4 − 2 = ____

____ more 🦋

2.

8 − 5 = ____

____ more 🐞

3.

7 − 3 = ____

____ more ⚽

4.

5 − 4 = ____

____ more ⛸

Explain It • Daily Reasoning

Why do you subtract to find how many more ⚪ than ⚫ there are?

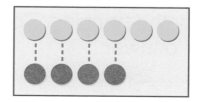

Chapter 4 • Using Subtraction

Practice and Problem Solving

Draw lines to match.
Subtract to find how many more.

1.

   6 − 3 = __3__

   __3__ more

2.

   4 − 1 = ____

   ____ more

3.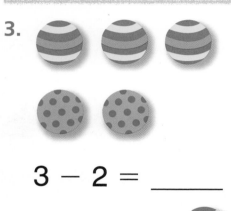

   3 − 2 = ____

   ____ more

4.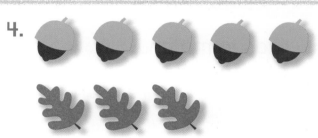

   5 − 3 = ____

   ____ more

Problem Solving
Application

Solve.

5. John had 7 stamps.
   Sue had 3 stamps.
   How many more stamps did John have?

   ____ stamps

6. Jane had 5 pennies.
   Mark had 2 pennies.
   How many more pennies did Jane have?

   ____ pennies

 Write About It • Look at Exercise 6. Draw the pennies. Then compare to check your answer.

🏠 HOME ACTIVITY • Have your child make up subtraction stories for you to solve.

56 fifty-six

Name _____

Problem Solving Strategy
Draw a Picture

Kathy had a box of 8 crayons.
She gave some crayons away.
She has 5 left.
How many crayons did Kathy give away?

UNDERSTAND

What do you want to find out?

Draw a line under the question.

PLAN

You can draw a picture to solve the problem.

SOLVE

    __3__ crayons

What number do I add to 5 to get 8?
$8 - \blacksquare = 5$
$5 + \underline{3} = 8$

CHECK

Does your answer make sense?
Explain.

Draw a picture to solve the problem.
Write how many were given away.

What number do I add to 3 to make 10?

1. I had 10 pencils.
   I gave some away.
   I have 3 left. How many
   pencils did I give away?

   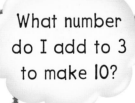

   _____ pencils

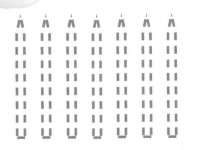

Chapter 4 • Using Subtraction

Problem Solving Practice

Keep in Mind!
Understand
Plan
Solve
Check

Draw a picture to solve the problem.
Write how many were given away.

1. Pat had 9 markers.
She gave some away.
She has 4 left.
How many markers did Pat give away?

What number do I add to 4 to make 9?

 markers

2. Dave had 8 stickers.
He gave some to Joe.
He has 6 left.
How many stickers did Dave give to Joe?

What number do I add to 6 to make 8?

_____ stickers

3. Ed had 3 erasers.
He gave some away.
He has 1 left.
How many erasers did Ed give away?

What number do I add to 1 to make 3?

_____ erasers

HOME ACTIVITY • Tell a math story like those in the problems on this page. Ask your child to draw a picture to solve the problem and then explain the drawing to you.

Name _____

Extra Practice

Complete the subtraction sentences.
Show ways to subtract from 7.

1. 7 – ____ = ____

2. 7 – ____ = ____

Show ways to subtract from 9.

3. 9 – ____ = ____

4. 9 – ____ = ____

Write the difference.

5. 5
  −0

6. 8
  −8

7. 7
  −2

8. 10
  − 7

9. 9
  −5

Draw lines to match.
Subtract to find how many more.

10.

4 – 3 = ____

____ more

11.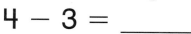
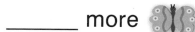

6 – 2 = ____

____ more

Problem Solving

Draw a picture to solve the problem.

12. Elise had 10 pencils. She gave some away. She has 6 left. How many pencils did Elise give away?

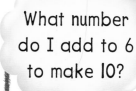

____ pencils

What number do I add to 6 to make 10?

Chapter 4 • Using Subtraction

Name _____

✓ Review/Test

Concepts and Skills

Complete the subtraction sentences.
Show ways to subtract from 8.

1. 8 − ____ = ____

2. 8 − ____ = ____

Show ways to subtract from 10.

3. 10 − ____ = ____

4. 10 − ____ = ____

Write the difference.

5. 6
   −4
   ___

6. 9
   −6
   ___

7. 8
   −1
   ___

8. 10
   − 9
   ___

9. 3
   −3
   ___

Draw lines to match.
Subtract to find how many more.

10.

6 − 3 = ____

____ more

11.

5 − 2 = ____

____ more

Problem Solving

Draw a picture to solve the problem.

12. Fran had 5 erasers. She gave some to a friend. She has 3 left. How many erasers did Fran give to her friend?

____ erasers

What number do I add to 3 to make 5?

60 sixty

Name _____

⭐Standardized Test Prep
Chapters 1–4

Choose the answer for questions 1–5.

1. Which subtraction sentence tells how many more cats than pandas there are?

 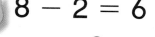

2. Which is another way to write 6 − 2 = 4?

$$\begin{array}{r}8\\-2\\\hline 6\end{array} \qquad \begin{array}{r}8\\-6\\\hline 2\end{array} \qquad \begin{array}{r}4\\-2\\\hline 2\end{array} \qquad \begin{array}{r}6\\-2\\\hline 4\end{array}$$
○ ○ ○ ○

3. 8 − 2 = ____

| 12 | 11 | 6 | 5 |
| ○ | ○ | ○ | ○ |

4. 9 − 3 = ____

| 6 | 7 | 12 | 13 |
| ○ | ○ | ○ | ○ |

5. Ann has 4 flowers. She gives 2 away. How many are left?

| 1 | 2 | 4 | 6 |
| ○ | ○ | ○ | ○ |

Show What You Know

6. Write a subtraction sentence that equals 10. Use to explain your answer. Draw the you use.

____ ◯ ____ ◯ ____

Chapter 4

sixty-one **61**

IT'S IN THE BAG
Cat Pocket-Dots

PROJECT You will make a cat pocket to hold your dot cards to practice your math facts.

You Will Need

- Lunch-size bag
- Blackline patterns
- Crayons
- Glue
- Scissors

Directions

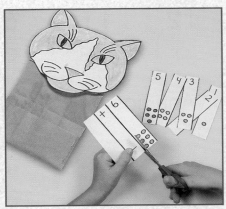

1. Color the cat face. Then cut it out.

2. Lay the bag in front of you. Put the flap at the top facing you. Glue the cat face to the flap.

3. Color the dots on the dot cards. Cut the cards apart.

4. Use dot cards 1 to 6 to make number sentences.

My Cat Biff

written by
Cheryl Michaels
illustrated by
Diane Greenseid

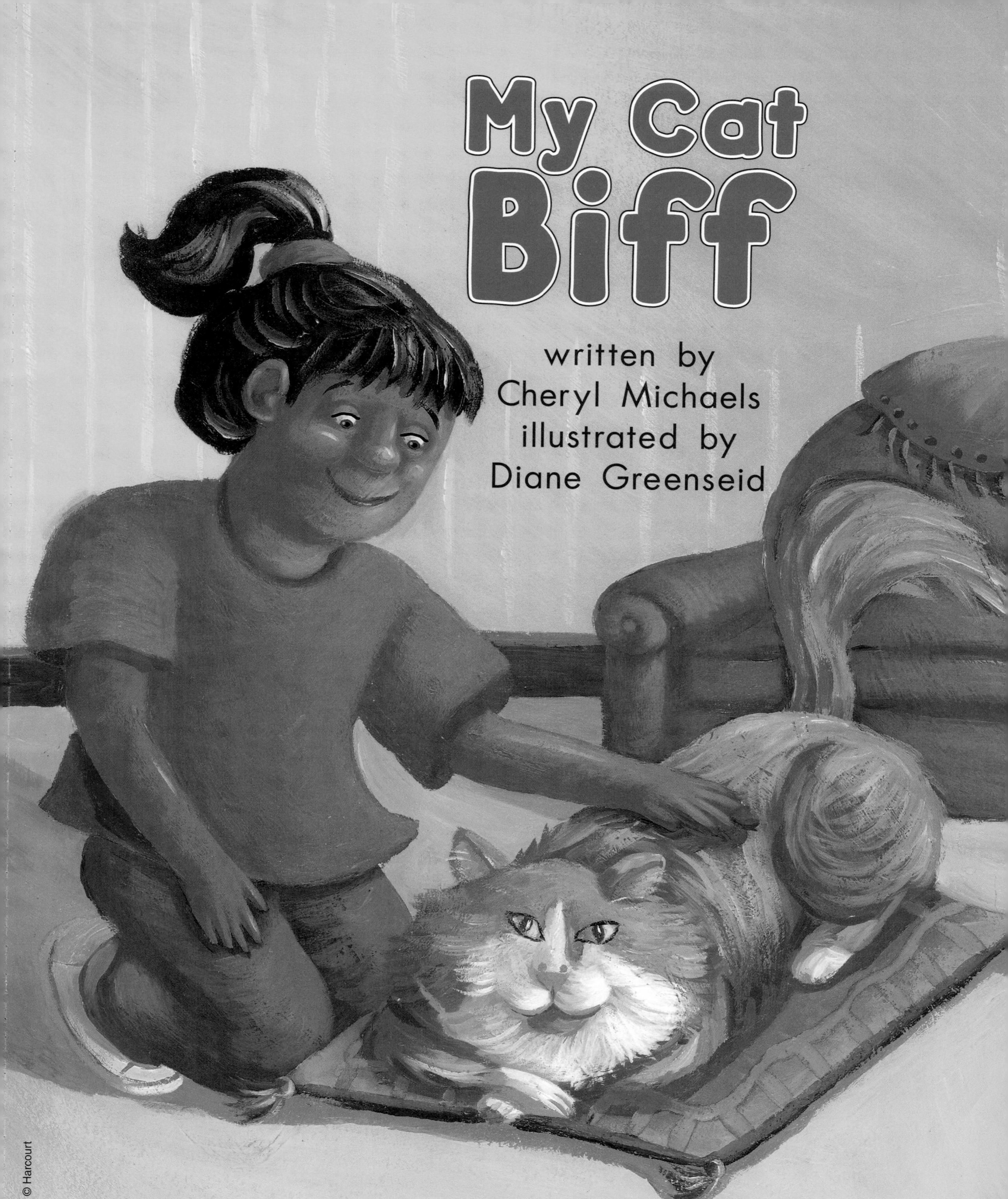

🔷 This book will help me review addition and subtraction stories.

This book belongs to _____.

Biff had 3 balls.
I gave him one more.

3 + 1 = _____

Biff had 4 balls.
I gave him two more.

4 + 2 = ___

Biff had 6 balls.
He hit one under the chair.

6 − 1 = ____

Name _____

PROBLEM SOLVING ON LOCATION

At the Petting Zoo

Old MacDonald's Petting Zoo is near Huntsville, Alabama. You can see and pet animals.

Use the pictures to help you write the addition sentence.

1 How many in all?

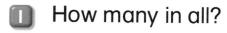

_____ ◯ _____ = _____ animals

2 How many in all?

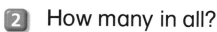

_____ ◯ _____ = _____ animals

llama

3 How many in all?

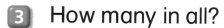

_____ ◯ _____ = _____ animals

emu

Unit 1 • Chapters 1–4

sixty-three **63**

Name _____

CHALLENGE

Equals

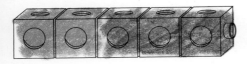 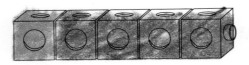

Both sides show 5. They are equal.

__2__ + __3__ = __1__ + __4__

5 = 5

Use different numbers of and ■.
Show sides that are equal.
Color. Write the numbers.

1. 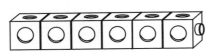 =

   ___ + ___ = ___ + ___

2. 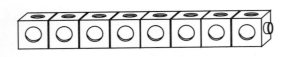 =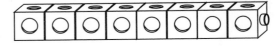

   ___ + ___ = ___ + ___

3.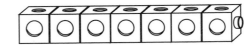

   ___ + ___ = ___ + ___

4.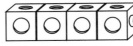

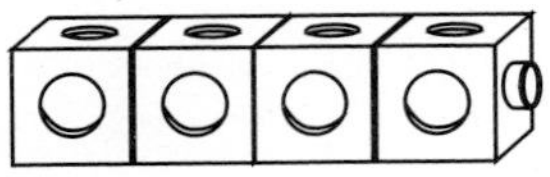

   ___ + ___ = ___ + ___

Name _____

✓ Study Guide and Review

Vocabulary

Add. Circle the **sum**.

Subtract. Circle the **difference**.

1.

5 + 1 = ____

2.

2 − 0 = ____

Skills and Concepts

Draw circles to show each number.
Write the sum.

3.

0 + 5 = ____

4.

2 + 2 = ____

Cross out pictures to subtract. Write the difference.

5.

5 − 2 = ____

6.

6 − 5 = ____

Write the subtraction sentence.

7.

____ ◯ ____ ◯ ____

8.

____ ◯ ____ ◯ ____

Unit 1 • Study Guide and Review

sixty-five **65**

Complete the addition or subtraction sentence.

9. Show a way to subtract from 9.

9 – ____ = ____

10. Show a way to make 10.

____ + ____ = 10

11. Write the numbers to match the dots. Write the sum.

____ + ____ = ____

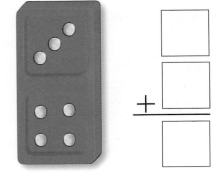

Draw lines to match.
Subtract to find how many more.

12.

6 – 2 = ____

____ more

13.

5 – 4 = ____

____ more

Problem Solving

Draw a picture. Then write an addition sentence to solve.

14. 2 fish swim.
1 more fish comes.
How many fish are there in all?

____ fish

Performance Assessment

Pick Up Jacks

Ann and Jenny were playing with jacks.

- Jenny placed some jacks in a circle.
- Ann picked up 2 jacks from the circle.
- After Ann picked up the 2 jacks, there were fewer than 4 jacks in the circle.

Write a subtraction number sentence that fits this math story. Draw pictures to help you.

Show your work.

Unit 1 • Performance Assessment

Name _____

TECHNOLOGY

The Learning Site • Seashell Search

1. Go to www.harcourtschool.com.
2. Click on [bucket icon].
3. Start. [arrow icon]
4. Add to play.

Practice and Problem Solving

Write the sum or difference.

1. $2 + 4 =$ ____
2. $3 + 1 =$ ____
3. $6 + 4 =$ ____

4. $6 - 5 =$ ____
5. $5 - 1 =$ ____
6. $7 - 3 =$ ____

Write two ways to make 7.

7. ____ ◯ ____ ◯ ____
8. ____ ◯ ____ ◯ ____

9. Sara has 2 shells. She finds 4 more. How many shells does she have in all?

   _____ shells

10. Jim has 9 pennies. He gives 2 away. How many pennies does he have left?

    _____ pennies

TECHNOLOGY

68 sixty-eight

LOOKING BACK
SCHOOL HOME CONNECTION

Dear Family,

In Unit 1 we learned how to find sums and differences. Here is a game for us to play together. This game will give me a chance to share what I have learned.

Love,

_____

Directions
1. Put a bean on a space.
2. Use pennies to show one way to make that number.
3. Take turns. Your partner uses rocks instead of beans.
4. If you get the same number again, show a new way to make it.
5. The first person to cover 3 spaces in a row wins.

Materials
- 8 beans (or other small objects)
- 8 rocks (or other small objects)
- 10 pennies

Unit 2 • Unit Game

LOOKING FORWARD
SCHOOL HOME CONNECTION

Dear Family,

During the next few weeks, we will learn different ways to memorize addition and subtraction facts. Here is important math vocabulary and a list of books to share.

Love,

_____

Vocabulary
- count on
- count back
- fact family

Vocabulary Power

count on A way to add by counting on from the greater number.

$$6 + 2 = 8$$

Say 6. Count on 2.
7, 8

count back A way to subtract by counting backward from the greater number.

$$5 - 2 = 3$$

Say 5. Count back 2.
4, 3

A **fact family** includes all the addition and subtraction facts that use the same numbers.

$$5 + 2 = 7 \qquad 7 - 2 = 5$$
$$2 + 5 = 7 \qquad 7 - 5 = 2$$

BOOKS TO SHARE

To read about addition and subtraction with your child, look for these books in your library.

Domino Addition, by Lynette Long Ph.D., Charlesbridge, 1996.

How Many, How Many, How Many, by Rick Walton, Candlewick, 1996.

Ten Sly Piranhas, by William Wise, Penguin Putnam, 1993.

Seven Little Rabbits, by John Becker, Walker, 1994.

Visit *The Learning Site* for additional ideas and activities. www.harcourtschool.com

CHAPTER 5 Addition Strategies

FUN FACTS

An octopus has a soft, bag-shaped body and 8 rubbery arms.

Theme: Sea Life

Name _____

✓ Check What You Know

Addition Patterns

Count the 🎈. Draw one more.
Write how many in all.

1.  🎈

   1 + 1 = ____

2.  🎈🎈

   2 + 1 = ____

3. 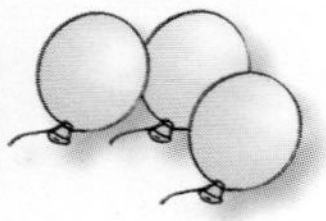 🎈🎈🎈

   3 + 1 = ____

4.  🎈🎈🎈🎈

   4 + 1 = ____

5.  🎈🎈🎈🎈🎈

   5 + 1 = ____

6.  🎈🎈🎈🎈🎈

   6 + 1 = ____

7.  🎈🎈🎈🎈🎈🎈

   7 + 1 = ____

Name _____

Count On 1 and 2

Vocabulary
count on

Explore

Say 5. Count on 1.

$5 + 1 = \underline{6}$

Say 6. Count on 2.

$6 + 2 = \underline{8}$

Connect

Use ●. Count on. Write the sum.

1.

   $7 + 2 = \underline{}$

2.

   $8 + 1 = \underline{}$

3.

   $4 + 2 = \underline{}$

4.

   $8 + 2 = \underline{}$

Explain It • Daily Reasoning

Which number would you start with to find the sum for 2 + 6? Does it matter? Explain.

$2 + 6 = ?$

Chapter 5 • Addition Strategies

Practice and Problem Solving

Use ●. Count on. Write the sum.

1.
   4 + 1 = 5

2.
   4 + 2 = ___

3. 7 + 2 = ___

4. 5 + 1 = ___

5. 5 + 2 = ___

6. 7 + 1 = ___

7. 3 + 2 = ___

8. 8 + 1 = ___

9. 2 + 1 = ___

10. 8 + 2 = ___

11. 3 + 1 = ___

Problem Solving
Algebra

Complete the addition sentence.

12. 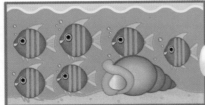 8 in all

    6 + ■ = 8
    ■ = ___

13. 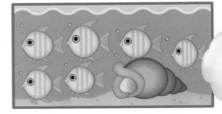 7 in all

    6 + ■ = 7
    ■ = ___

Write About It • Look at Exercise 13. Explain how you got your answer.

🏠 **HOME ACTIVITY** • Choose numbers from 1 to 7, and have your child count on 1 or 2.

72 seventy-two

CHAPTER 6
Addition Facts Practice

FUN FACTS

Children who like to exercise usually stay active from year to year.

Theme: On the Playground

Name _____

✓ Check What You Know

Add in Any Order

Add. Circle the addition sentences in each row that have the same sum.

1. 3 + 1 = ___   2. 1 + 3 = ___   3. 1 + 2 = ___

4. 2 + 5 = ___   5. 7 + 1 = ___   6. 5 + 2 = ___

7. 4 + 2 = ___   8. 4 + 1 = ___   9. 2 + 4 = ___

10. 0 + 3 = ___   11. 1 + 3 = ___   12. 3 + 0 = ___

Use a Number Line to Count On

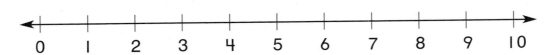

Use the number line. Count on to find the sum.

13. 4   14. 3   15. 5   16. 7   17. 6
    +2   +2   +1   +2   +2

18. 8   19. 4   20. 6   21. 2   22. 3
    +2   +3   +3   +1   +1

Name _____

Use the Strategies

Learn

What helps you remember these facts?

1 + 0 = 1
1 + 1 = 2
1 + 2 = 3
1 + 3 = 4

I use add 0, doubles, or count on.

Check

Add. Write the sums.

1. **Count On 1**
   5 + 1 = 6
   6 + 1 = ___
   7 + 1 = ___
   8 + 1 = ___
   9 + 1 = ___

2. **Count On 2**
   4 + 2 = ___
   5 + 2 = ___
   6 + 2 = ___
   7 + 2 = ___
   8 + 2 = ___

3. **Count On 3**
   4 + 3 = ___
   5 + 3 = ___
   6 + 3 = ___
   7 + 3 = ___

4. **Add 0**
   6 + 0 = ___
   7 + 0 = ___
   8 + 0 = ___
   9 + 0 = ___
   10 + 0 = ___

5. **Use Doubles**
   1 + 1 = ___
   2 + 2 = ___
   3 + 3 = ___
   4 + 4 = ___
   5 + 5 = ___

Explain It • Daily Reasoning

Why is 3 + 3 = 6 a doubles fact?

Chapter 6 • Addition Facts Practice

eighty-five **85**

Practice and Problem Solving

Add. Write the sum.

1. 4 +3 = 7
2. 7 +3
3. 5 +2
4. 8 +1
5. 2 +2

6. 5 +3
7. 5 +1
8. 6 +1
9. 4 +4
10. 9 +1

11. 4 +2
12. 4 +5
13. 3 +3
14. 1 +1
15. 3 +2

16. 6 +3
17. 5 +5
18. 9 +0
19. 6 +2
20. 8 +2

Problem Solving
Logical Reasoning

21. There are 9 children in all. 2 are outside the playhouse, and the rest are inside. How many children are inside?

_____ children are inside.

 Write About It • Look at Exercise 21. Draw pictures to show how you got your answer.

HOME ACTIVITY • On each day of the week, choose a different number and work with your child to practice the facts that have that sum. For example, on Monday, practice all the facts that have a sum of 5.

86 eighty-six

Name _____

Sums to 8

Learn

You can change the order of the numbers you add. The sum is the same.

$$\begin{array}{r} 5 \\ +3 \\ \hline 8 \end{array}$$

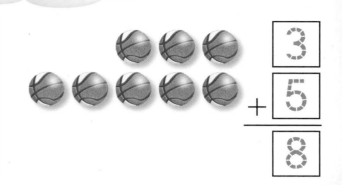

$$\begin{array}{r} 3 \\ +5 \\ \hline 8 \end{array}$$

Check

Add. Change the order.
Write the new fact.

1. $\begin{array}{r} 5 \\ +2 \\ \hline \end{array}$

2. $\begin{array}{r} 4 \\ +3 \\ \hline \end{array}$

3. $\begin{array}{r} 1 \\ +7 \\ \hline \end{array}$

4. $\begin{array}{r} 6 \\ +2 \\ \hline \end{array}$

5. $\begin{array}{r} 7 \\ +0 \\ \hline \end{array}$

6. $\begin{array}{r} 5 \\ +1 \\ \hline \end{array}$

Explain It • Daily Reasoning

Does the sum change when you change the order of the numbers you are adding? Explain.

Practice and Problem Solving

1. Add. Use the key. Color each space by the sum. What patterns do you see?

2 +3	5 +2	0 +5	3 +4	4 +1
3 +3	2 +6	6 +0	3 +5	2 +4
7 +0	1 +4	1 +6	3 +2	2 +5
6 +2	1 +5	7 +1	4 +2	0 +8

Problem Solving
Mental Math

Solve. Draw a picture to check.

2. Eric has 8 flowers in two pots. He has the same number in each pot. How many flowers are in each pot?

_____ flowers

 Write About It • Write a math story about this number sentence. Draw a picture to show your story.

HOME ACTIVITY • Ask your child to see how many addition facts with a sum of 8 he or she can write.

Name _____

Sums to 10

Learn

The order of the numbers changed. The sum is the same.

```
  5    4
 +4   +5
 ──   ──
  9    9
```

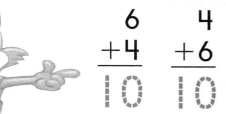

```
  6    4
 +4   +6
 ──   ──
 10   10
```

Check

Add. Write the sums.

1. 9   0
   +0   +9

2. 9   1
   +1   +9

3. 7   2
   +2   +7

4. 7   3
   +3   +7

5. 5   4
   +4   +5

6. 4   3
   +3   +4

7. 10   0
   + 0   +10

8. 8   1
   +1   +8

9. 6   2
   +2   +6

10. 6   3
   +3   +6

11. 3   5
   +5   +3

12. 5   2
   +2   +5

13. 8   2
   +2   +8

14. 4   2
   +2   +4

15. 1   7
   +7   +1

Explain It • Daily Reasoning

When is the sum the same as one of the two numbers you are adding?

Chapter 6 • Addition Facts Practice

Practice and Problem Solving

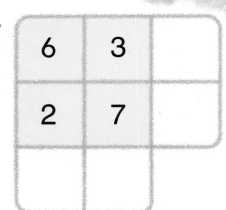

Add across. Add down.
Write the sums.

1.
2	4	6
5	3	8
7	7	

2.
6	3	
2	7	

3.
4	6	
4	0	

4.
8	2	
1	2	

Problem Solving
Mental Math

Circle two ways to name the same number.

5. $6 + 2$ $4 + 4$ $5 + 1$ $3 + 2$

6. $0 + 9$ $7 + 3$ $6 + 3$ $5 + 2$

 Write About It • Look at Exercise 6. Write another way to name the same number.

HOME ACTIVITY • Say a number from 1 to 10, and ask your child to tell you an addition fact that has that number as its sum. Repeat the activity for a new sum.

Name _____

Algebra: Follow the Rule

Vocabulary
rule

Learn

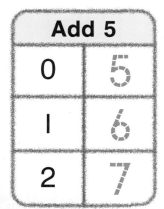

Add 5	
0	5
1	6
2	7

The rule is Add 5. I add 5 each time.

Check

Complete the table. Follow the rule.

1.
Add 2	
4	
6	
8	

2.
Add 3	
4	
3	
2	

3.
Add 4	
4	
5	
6	

4.
Add 1	
9	
8	
7	

5.
Add 2	
1	
2	
3	

6.
Add 0	
8	
9	
10	

Explain It • Daily Reasoning

Does every table on this page have a pattern? Explain.

Chapter 6 • Addition Facts Practice

Practice and Problem Solving

Complete the table. Follow the rule.

1. **Add 3**

5	8
6	9
7	10

2. **Add 5**

5	
4	
3	

3. **Add 1**

4	
5	
6	

4. **Add 2**

3	
5	
7	

5. **Add 4**

3	
2	
1	

6. **Add 6**

3	
2	
1	

Problem Solving
Logical Reasoning

Write the rule.

7. **Add ___**

2	5
4	7
6	9

8. **Add ___**

4	9
2	7
0	5

9. **Add ___**

7	7
8	8
9	9

Write About It • Look at Exercise 9. Explain how you figured out the rule.

HOME ACTIVITY • Ask your child to write an addition rule. Help your child make a table that follows the rule.

Name _____

Problem Solving Strategy
Write a Number Sentence

<u>6 children play.</u>

<u>3 more come.</u>

How many children are there now?

UNDERSTAND

What information do you know?
Underline it.

PLAN

How can you solve this problem?
Write a number sentence.

SOLVE

__6__ children play

__3__ more come

__6__ ⊕ __3__ ⊖ __9__
children

CHECK

Does your answer make sense?
Explain.

Solve. Write a number sentence.
Draw a picture to check.

1. 3 boys go down the slide.
   3 girls go down the slide.
   How many children in all go down the slide?

   _____ ◯ _____ ◯ _____
   children

   THINK:
   How can I find out how many children there are in all?

Chapter 6 • Addition Facts Practice

ninety-three **93**

Problem Solving Practice

Keep in Mind!
Understand
Plan
Solve
Check

Solve. Write a number sentence.
Draw a picture to check.

THINK: What do I need to use to solve the problem?

1. There are 8 bean bags. Kendra finds 1 more. How many bean bags are there in all?

   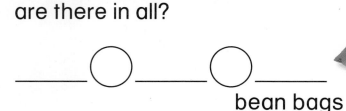
   ____ ◯ ____ ◯ ____
   bean bags

2. 6 boys run races. 2 girls join them. How many children run in all?

   ____ ◯ ____ ◯ ____
   children

3. There are 4 soccer balls in one bag. There are 6 in another bag. How many soccer balls are there in all?

   ____ ◯ ____ ◯ ____
   soccer balls

4. 3 girls jump rope. 4 more girls join them. How many girls jump rope?

   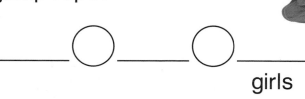
   ____ ◯ ____ ◯ ____
   girls

HOME ACTIVITY • Tell your child a math story, and ask him or her to write a number sentence to solve it. For example: "A boy has 5 toy cars. Then he finds 3 more. How many toy cars does he have now?" (5 + 3 = 8)

Name _____

Extra Practice

Complete the table. Follow the rule.

Add. Change the order. Write the new fact.

1.
Add 1	
9	
8	
7	

2. 9
  +0

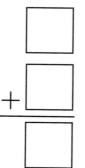

3. 6
  +2

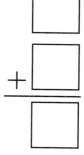

Add. Write the sum.

4. 4
  +2

5. 6
  +3

6. 1
  +1

7. 8
  +2

8. 7
  +2

9. 5
  +3

10. 5
  +1

11. 4
  +4

12. 4
  +3

13. 6
  +1

14. 7
  +3

15. 2
  +2

Problem Solving

Solve. Write a number sentence. Draw a picture to check.

16. 2 girls are playing catch. 4 more girls join them. How many girls are there in all?

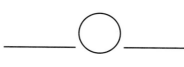 
girls

Chapter 6 • Addition Facts Practice

Name _____

✓ Review/Test

Concepts and Skills

Complete the table. Follow the rule.

1.
Add 3	
4	
5	
6	

Add. Change the order. Write the new fact.

2. 2
   +8
   ___
   □

   □
  +□
   ___
   □

3. 4
   +5
   ___
   □

   □
  +□
   ___
   □

Add. Write the sum.

4. 9   5. 5   6. 4   7. 7   8. 6   9. 4
   +1   +2   +4   +2   +3   +2

10. 3   11. 7   12. 6   13. 8   14. 5   15. 7
   +3   +1   +2   +1   +5   +3

Problem Solving

Solve. Write a number sentence. Draw a picture to check.

16. 5 boys play kickball. 4 more boys join them. How many boys are playing in all?

boys

96 ninety-six

Name _____

★Standardized Test Prep
Chapters 1-6

Choose the answer for questions 1-6.

1. 2 + 3 = ____

1	5	7	8
○	○	○	○

2. 7 + 1 = ____

6	8	10	13
○	○	○	○

3. 5 + 2 = ____

1	2	7	9
○	○	○	○

4. 5 + 5 = ____

0	6	10	16
○	○	○	○

5. Which is a way to make 10?

7 + 2	6 + 1	5 + 3	8 + 2
○	○	○	○

6. Kathy has 9 goldfish. She buys 1 more. How many goldfish does she have in all?

12	10	7	6
○	○	○	○

Show What You Know

7. Write a rule.
   Write numbers in the table that explain the rule.

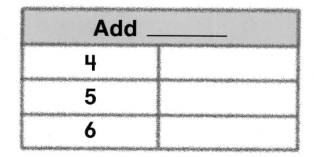

Chapter 6

ninety-seven **97**

Problem Solving Strategy
Draw a Picture

6 ladybugs were flying.

Some landed on a leaf.

4 ladybugs are still in the air.

How many ladybugs landed?

UNDERSTAND

What do you want to find out?

Draw a line under the question.

PLAN

You can draw a picture to solve the problem.

SOLVE

6 − ☐ = 4

What number do I add to 4 to get 6?

4 + __2__ = 6

__2__ ladybugs landed.

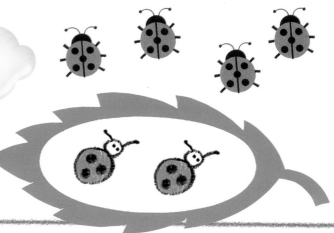

CHECK

Does your answer make sense? Explain.

Draw a picture to solve the problem.

What number do I add to 5 to make 7?

1. 7 birds were flying.
   Some landed on a tree.
   5 birds are still in the air.
   How many birds landed?

   _____ birds

Chapter 7 • Subtraction Strategies

Problem Solving Practice

Keep in Mind!
Understand
Plan
Solve
Check

Draw a picture to solve the problem.

1. 9 butterflies were flying. Some butterflies landed on a bush. 6 butterflies are still in the air. How many butterflies landed?

   What number do I add to 6 to make 9?

   _____ butterflies

2. 8 kites were flying. Some kites came down. 4 kites are still in the air. How many kites came down?

   What number do I add to 4 to make 8?

   _____ kites

3. 10 bees were flying. Some went into a hive. 7 bees are still in the air. How many bees went into the hive?

   What number do I add to 7 to make 10?

   _____ bees

HOME ACTIVITY • Tell a math story like these in the problems on this page. Ask your child to draw a picture to solve the problem and then explain his or her drawing to you.

Name _____

Extra Practice

Use the number line.
Count back to subtract.

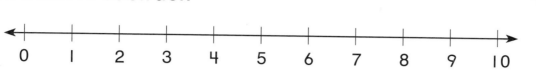

1. $4 - 1 = \underline{}$

2. $3 - 2 = \underline{}$

3. $7 - 3 = \underline{}$

4. $8 - 1 = \underline{}$

5. $2 - 2 = \underline{}$

6. $10 - 3 = \underline{}$

7. $5 - 1 = \underline{}$

8. $5 - 2 = \underline{}$

Add. Then subtract.

9.   6   8
    +2   −2

10.   5   6
    +1   −1

11.   4   6
    +2   −2

Problem Solving

Draw a picture to solve the problem.

12. 5 birds were flying.
Some landed on a tree.
2 are still in the air.
How many birds landed?

_____ birds

What number do I add to 2 to make 5?

Chapter 7 • Subtraction Strategies

Name _____

✓ Review/Test

Concepts and Skills

Use the number line.
Count back to subtract.

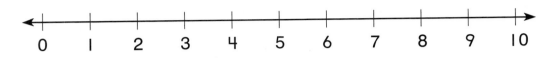

1. 7 − 1 = ___
2. 4 − 3 = ___
3. 8 − 2 = ___
4. 10 − 1 = ___
5. 3 − 3 = ___
6. 9 − 2 = ___
7. 6 − 2 = ___
8. 9 − 1 = ___

Add. Then subtract.

9.   7   10
     +3   − 3

10.   6   9
     +3   −3

11.   2   9
     +7   −7

Problem Solving

Draw a picture to solve the problem.

12. 8 bees were flying. Some went into a hive. 6 bees are still flying. How many bees went into the hive?

_____ bees

What number do I add to 6 to make 8?

110 one hundred ten

Name _____

★Standardized Test Prep
Chapters 1–7

Choose the answer for questions 1–4.

1. Which does the number line show?

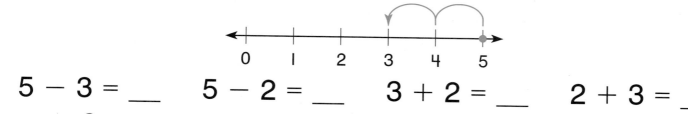

5 − 3 = __   5 − 2 = __   3 + 2 = __   2 + 3 = __
○   ○   ○   ○

2. Which fact is related to 8 + 2 = 10?

6 + 2 = 8   7 + 1 = 8   8 − 2 = 6   10 − 8 = 2
○   ○   ○   ○

3. Which does the number line show?

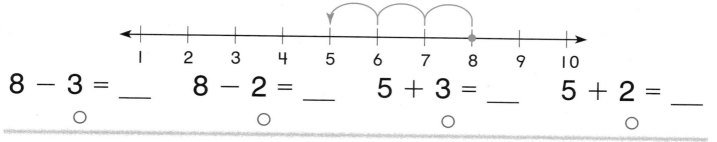

8 − 3 = __   8 − 2 = __   5 + 3 = __   5 + 2 = __
○   ○   ○   ○

4. Which does the number line show?

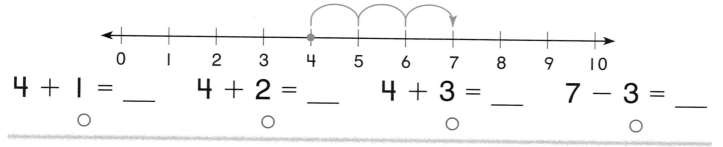

4 + 1 = __   4 + 2 = __   4 + 3 = __   7 − 3 = __
○   ○   ○   ○

Show What You Know

5. Write numbers on the number line.
   Draw arrows to show subtraction.
   Write the subtraction sentence to explain.

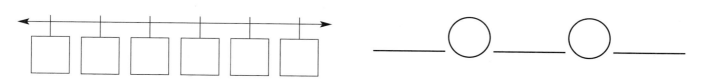

Name _____

MATH GAME

Up, Up, and Away

Play with a partner.

1. Put your 🎲 at START.
2. Stack the subtraction cards face down.
3. Take a card, and find the difference.
4. Move your 🎲 that many spaces.
5. The first player to get to END wins.

You will need
subtraction cards
2 🎲

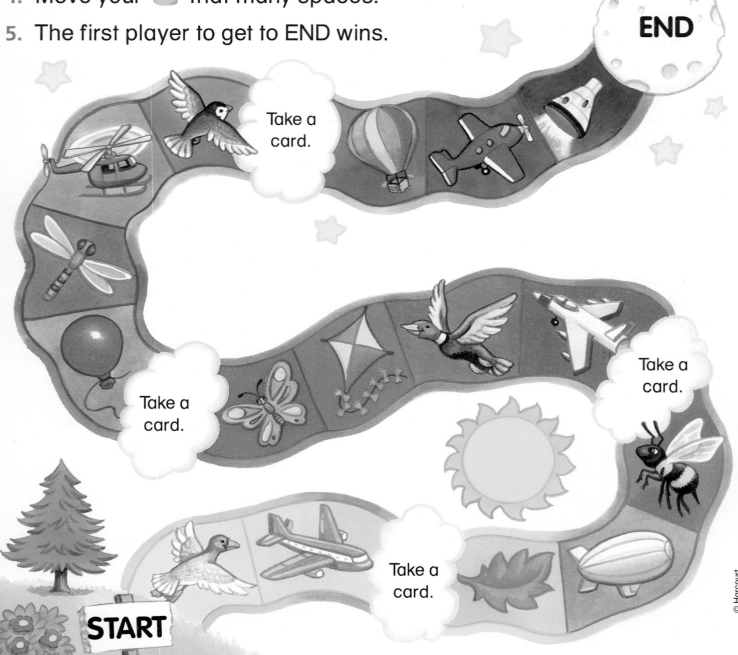

CHAPTER 8
Subtraction Facts Practice

FUN FACTS

The starfruit is star-shaped when cut. It has 6 points.

Theme: Fun Food

Name _____

✓ Check What You Know

Count Back to Subtract: Facts to 10

Use the number line. Count back to subtract.

1. [number line 0–5, dot at 4]
   $4 - 1 = \underline{}$

2. [number line 0–5, dot at 3]
   $3 - 2 = \underline{}$

3. [number line 0–5, dot at 5]
   $5 - 1 = \underline{}$

4. [number line 5–10, dot at 7]
   $7 - 2 = \underline{}$

5. [number line 5–10, dot at 8]
   $8 - 2 = \underline{}$

6. [number line 5–10, dot at 10]
   $10 - 1 = \underline{}$

Relate Addition and Subtraction

Add. Then subtract.

7.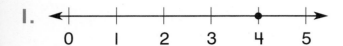

$6 + 3 = \underline{}$

$9 - 3 = \underline{}$

8.

$5 + 2 = \underline{}$

$7 - 5 = \underline{}$

9. $\begin{array}{r} 3 \\ +2 \\ \hline \end{array}$ $\begin{array}{r} 5 \\ -3 \\ \hline \end{array}$

10. $\begin{array}{r} 8 \\ +2 \\ \hline \end{array}$ $\begin{array}{r} 10 \\ -8 \\ \hline \end{array}$

11. $\begin{array}{r} 3 \\ +4 \\ \hline \end{array}$ $\begin{array}{r} 7 \\ -3 \\ \hline \end{array}$

Name _____

Use the Strategies

Learn

You can count back, subtract 0, or subtract all.

$$10 - 1$$
$$10 - 2$$
$$10 - 3$$
$$10 - 0$$
$$10 - 10$$

Check

Subtract. Write the difference.

1. **Count Back 1**

   $10 - 1 = \underline{9}$

   $9 - 1 = \underline{}$

   $8 - 1 = \underline{}$

   $7 - 1 = \underline{}$

2. **Count Back 2**

   $10 - 2 = \underline{8}$

   $9 - 2 = \underline{}$

   $8 - 2 = \underline{}$

   $7 - 2 = \underline{}$

3. **Count Back 3**

   $10 - 3 = \underline{7}$

   $9 - 3 = \underline{}$

   $8 - 3 = \underline{}$

   $7 - 3 = \underline{}$

4. **Subtract 0**

   $10 - 0 = \underline{10}$

   $9 - 0 = \underline{}$

   $8 - 0 = \underline{}$

5. **Subtract All**

   $10 - 10 = \underline{0}$

   $9 - 9 = \underline{}$

   $8 - 8 = \underline{}$

Explain It • Daily Reasoning

How can knowing that $10 - 3 = 7$ help you find the difference for $10 - 4$?

Chapter 8 • Subtraction Facts Practice

Practice and Problem Solving

Subtract. Circle the facts for **subtract 0** and for **subtract all**.

1. (9 − 9 = 0)
2. (8 − 0 = 8)
3. 7 − 2
4. 8 − 3
5. 6 − 3
6. 10 − 1

7. 7 − 3
8. 9 − 2
9. 10 − 2
10. 4 − 0
11. 6 − 1
12. 8 − 8

13. 7 − 1
14. 10 − 0
15. 9 − 3
16. 5 − 3
17. 4 − 2
18. 10 − 10

19. 6 − 0
20. 8 − 1
21. 7 − 7
22. 9 − 1
23. 8 − 2
24. 10 − 3

Problem Solving
Visual Thinking

25. Cross out some of the apples. Write a subtraction sentence to tell about the picture.

____ ◯ ____ ◯ ____

Write About It • Why is 0 the answer when you subtract all?

HOME ACTIVITY • Make subtraction flash cards with your child. Have your child find all the facts that have a difference of 1, for example, 4 − 3 = 1.

Name _____

Subtraction to 10

Learn

These facts use the same numbers.

```
  10        10
−  2      −  8
─────    ─────
   8         2
```

10 − 2 = 8, so
10 − 8 = 2.

Check

Subtract. Circle the pair of facts if they use the same numbers.

1. 9   9
   −1 −8
   ── ──
    8   1

2. 7   7
   −7 −0
   ── ──

3. 10 10
   −5 −9
   ── ──

4. 10 10
   −6 −4
   ── ──

5. 8   8
   −3 −7
   ── ──

6. 9   9
   −5 −4
   ── ──

7. 8   8
   −5 −4
   ── ──

8. 10 10
   −7 −3
   ── ──

9. 9   9
   −2 −7
   ── ──

10. 9   9
   −6 −3
   ── ──

11. 8   8
   −2 −6
   ── ──

12. 7   7
   −5 −1
   ── ──

Explain It • Daily Reasoning

If you know that 10 − 1 = 9, what other subtraction fact do you know? Explain.

10 − 1 = 9

Chapter 8 • Subtraction Facts Practice

one hundred seventeen 117

Practice and Problem Solving

Subtract across. Subtract down.

1.
5	4	1
3	2	1
2	2	0

2.
6	3	
4	3	

3.
7	1	
6	0	

4.
8	4	
7	4	

Problem Solving
Logical Reasoning

Solve the riddle. Write the number.

5. If you count back 2 from me, the answer is 3. What number am I?

   _____

6. If you count back 3 from me, the answer is 4. What number am I?

   _____

 Write About It • Look at Exercise 6. Explain how you got your answer.

HOME ACTIVITY • Have your child tell you all the subtraction facts from 8 − 0 through 8 − 8.

118 one hundred eighteen

Name _____

Algebra: Follow the Rule

Vocabulary
rule

Learn

Subtract 2

10	8
8	6
6	4

The rule is subtract 2, so I subtract 2 from each number.

Check

Complete the table. Follow the rule.

1. Subtract 1

7	
5	
3	

2. Subtract 5

10	
9	
8	

3. Subtract 3

5	
6	
7	

4. Subtract 0

5	
7	
9	

5. Subtract 4

7	
8	
9	

6. Subtract 2

2	
4	
6	

Explain It • Daily Reasoning

What patterns do you see? Explain.

Chapter 8 • Subtraction Facts Practice   one hundred nineteen **119**

Practice and Problem Solving

Complete the table. Follow the rule.

1. **Subtract 2**

5	3
4	
3	

2. **Subtract 0**

2	
4	
6	

3. **Subtract 3**

10	
9	
8	

4. **Subtract 1**

10	
8	
6	

5. **Subtract 5**

5	
6	
7	

6. **Subtract 4**

4	
5	
6	

Problem Solving

Logical Reasoning

Write the rule.

7. **Subtract _____**

7	5
5	3
3	1

8. **Subtract _____**

6	5
4	3
2	1

Write About It • Look at Exercise 8. Explain how you got your answer.

HOME ACTIVITY • Ask your child to write a subtraction rule and make a table that follows the rule. Have him or her explain how to use the table.

Name _____

Fact Families to 10

Vocabulary
fact family

Explore

4 + 2 = __6__

6 − 2 = __4__

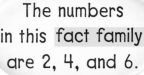
2 + 4 = __6__

6 − 4 = __2__

The numbers in this fact family are 2, 4, and 6.

Connect

Use 🟦 and 🟫 to add or subtract.
Write the numbers in the fact family.

1. 8 + 2 = ___
   2 + 8 = ___
   10 − 2 = ___
   10 − 8 = ___

   | 2 | 8 | 10 |

2. 4 + 1 = ___
   1 + 4 = ___
   5 − 1 = ___
   5 − 4 = ___

   | | | |

3. 7 + 2 = ___
   2 + 7 = ___
   9 − 2 = ___
   9 − 7 = ___

   | | | |

4. 8 + 1 = ___
   1 + 8 = ___
   9 − 1 = ___
   9 − 8 = ___

   | | | |

Explain It • Daily Reasoning

How many addition and subtraction facts are in the fact family for these numbers? Use 🟦 and 🟫 to prove your answer.

Chapter 8 • Subtraction Facts Practice

Practice and Problem Solving

Add or subtract. Write the numbers in the fact family.

1.

   6   2   8   8
   +2   +6   −2   −6
   ─   ─   ─   ─
   8   8   6   2

   2 6 8

2.

   4   3   7   7
   +3   +4   −3   −4
   ─   ─   ─   ─

   ☐ ☐ ☐

3.

   5   1   6   6
   +1   +5   −1   −5
   ─   ─   ─   ─

   ☐ ☐ ☐

Problem Solving
Algebra

4. Write the missing numbers.

   6   ☐   9   9
  +☐   +6   −☐   −6
   ─   ─   ─   ─
   9   9   6   ☐

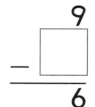

Write About It • Look at Exercise 4.
Explain why the number facts belong in a fact family.

▬ **HOME ACTIVITY** • Have your child write the facts in a fact family and then explain to you why those number sentences belong in that family.

122 one hundred twenty-two

Name _____

Problem Solving Skill
Choose the Operation

Jessie has 6 apples.
She gives 4 away.
How many apples does
she have left?

Some apples are taken away. I need to subtract.

add   (subtract)

__2__ apples

__6__ ⊖ __4__ ⊜ __2__

There are 3 forks on a table.
Ida brings 2 more.
How many are there now?

More forks are added. I need to add.

(add)   subtract

__5__ forks

__3__ ⊕ __2__ ⊜ __5__

Circle **add** or **subtract**.
Write the number sentence.

THINK: Do I add or subtract?

1. There are 7 pretzels.
   Children eat 4 of them.
   How many are left?

   add   subtract

   _____ pretzels

   ___ ◯ ___ ◯ ___

2. There are 9 carrots.
   Bunnies eat 5 of them.
   How many carrots
   are left?

   add   subtract

   _____ carrots

   ___ ◯ ___ ◯ ___

Chapter 8 • Subtraction Facts Practice

one hundred twenty-three **123**

Problem Solving Practice

Circle **add** or **subtract**.
Write the number sentence.

THINK: Do I add or subtract?

1. There are 5 muffins.
   Zoe brings 2 more.
   How many are there now?

   __7__ muffins

   (add)   subtract

   __5__ ⊕ __2__ ⊜ __7__

2. There are 6 apples.
   Children eat 2 of them.
   How many are left?

   _____ apples
   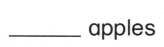

   add   subtract
   ___◯___◯___

3. There are 7 pears.
   John brings 3 more.
   How many are there now?

   _____ pears

   add   subtract
   ___◯___◯___

4. There are 8 sandwiches.
   Children eat 3 of them.
   How many are left?

   _____ sandwiches

   add   subtract
   ___◯___◯___

5. There are 8 oranges.
   Children eat 4 of them.
   How many are left?

   _____ oranges

   add   subtract
   ___◯___◯___

🏠 **HOME ACTIVITY** • For each problem, ask your child to tell how he or she decided whether to add or subtract.

Name _____

Extra Practice

Subtract. Circle the pair of facts if they use the same numbers.

1.
   $$6 - 2$$ $$6 - 4$$

2.
   $$7 - 5$$ $$7 - 3$$

3.
   $$8 - 3$$ $$8 - 5$$

Add or subtract.
Write the numbers in the fact family.

4.

   $$3 + 2$$ $$2 + 3$$ $$5 - 3$$ $$5 - 2$$

Complete the table. Follow the rule.

5.
Subtract 1	
9	
8	
7	

6.
Subtract 0	
5	
4	
3	

7.
Subtract 3	
3	
4	
5	

Problem Solving

Circle **add** or **subtract**.
Write the number sentence.

8. There are 6 acorns under a tree.
   A squirrel eats 3 of them.
   How many acorns are left?

   _____ acorns

   add   subtract

   _____ ◯ _____ ◯ _____

Chapter 8 • Subtraction Facts Practice   one hundred twenty-five **125**

Name _____

✓ Review/Test

Concepts and Skills

Subtract. Circle the pair of facts if they use the same numbers.

1. 9   9   2. 10   10   3. 7   7
   −5   −4   − 8   − 1   −4   −3

Add or subtract.
Write the numbers in the fact family.

4.

    4   2   6   6
   +2   +4   −2   −4

Complete the table. Follow the rule.

5. | Subtract 3 | |
   |---|---|
   | 10 | |
   | 9 | |
   | 8 | |

6. | Subtract 2 | |
   |---|---|
   | 8 | |
   | 9 | |
   | 10 | |

7. | Subtract 5 | |
   |---|---|
   | 5 | |
   | 7 | |
   | 9 | |

Problem Solving

Circle **add** or **subtract**.
Write the number sentence.

8. There are 10 grapes. Meg eats 2 of them. How many are left?

   _____ grapes

   add   subtract

   ____ ○ ____ ○ ____

Name _____

Standardized Test Prep
Chapters 1–8

Choose the answer for questions 1–6.

1. Which number completes the table?

Count Back 2
8 − 2 = 6
9 − 2 = 7
10 − 2 = ?

  4   6   8   10
○   ○   ○   ○

2. Which number completes the table?

Subtract 4	
4	0
5	1
6	?

  1   2   9   10
○   ○   ○   ○

3.   10 − 1 = ____

  9   10   11   13
○   ○   ○   ○

4.   9 − 2 = ____

  6   7   10   11
○   ○   ○   ○

5. Which subtraction fact is related to 4 + 3 = 7?

  7 − 5 = 2   8 − 4 = 4   7 − 3 = 4   10 − 3 = 7
     ○   ○   ○   ○

6. 3 pickles are in a bowl. 2 more are added. Which number sentence tells how many pickles are in the bowl now?

  3 − 2 = 1   3 − 3 = 0   3 + 2 = 5   2 + 2 = 4
     ○   ○   ○   ○

Show What You Know

7. Use 3, 7, and 10. Write the four number sentences that explain the fact family.

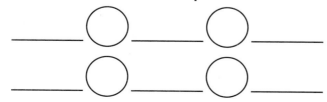 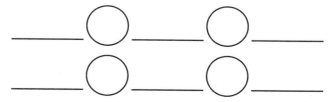

IT'S IN THE BAG
Math Under the Sea

PROJECT Create your own snorkel mask to practice your math facts.

You Will Need
- Blackline patterns
- Crayons
- Scissors

Directions

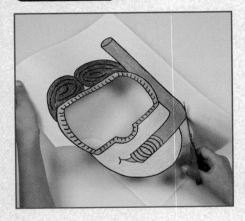

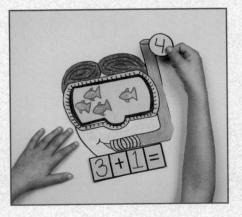

1. Color the snorkel person.
2. Cut out the snorkel person.
3. Draw an addition or subtraction problem. Use turtles, fish, or shells in your problem.
4. Write the addition or subtraction sentence. Then write the answer on a bubble.
5. Draw other problems. Write the number sentences and answers.

UNDER THE Sea

written by Jo Sumara
illustrated by Jui Ishida

This book will help me review addition and subtraction facts.

This book belongs to _____.

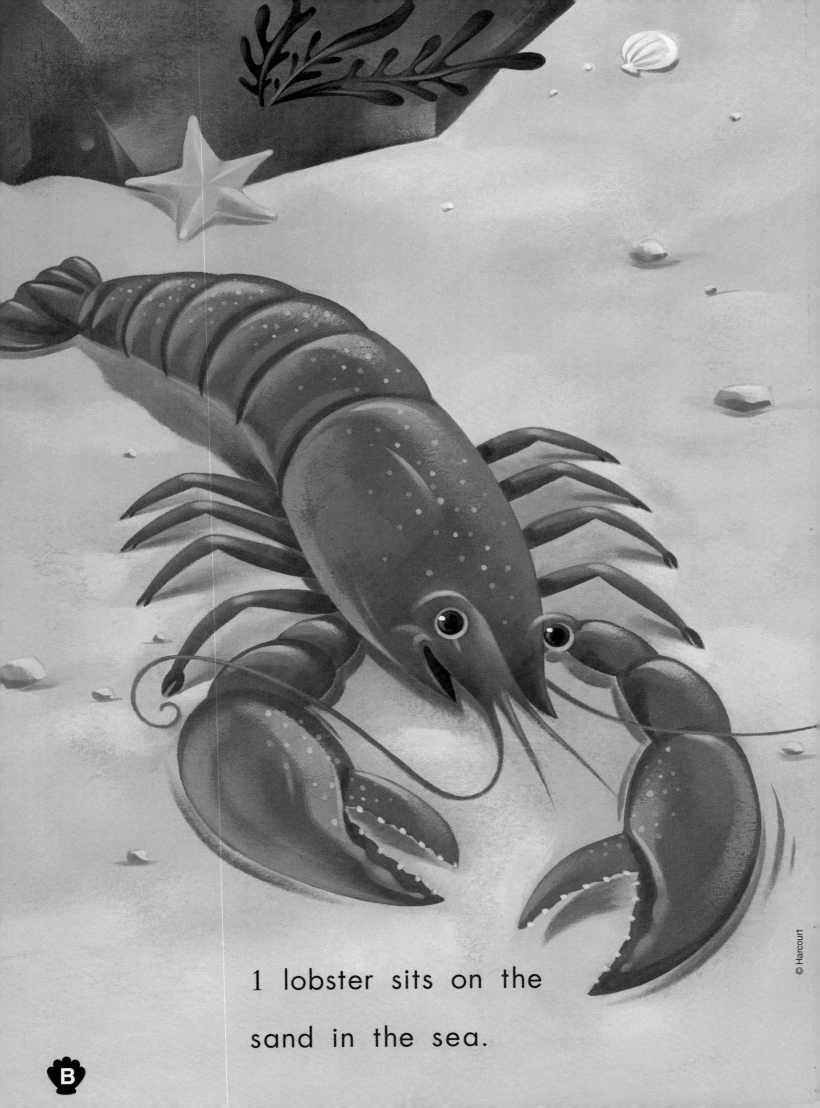

1 lobster sits on the sand in the sea.

2 more come along,
so now there are ____.

6 dolphins jump up,
do a flip, and then dive.

1 swims away,

so now there are ____.

1 sea horse floats by.
He looks like a hero.

Soon he is gone,
and then there are ____.

Name _____

PROBLEM SOLVING ON LOCATION

At the Aquarium

You can see clown fish and sea horses at the Tennessee Aquarium.

Each tank has and .

Draw a picture and write a number sentence to solve.

Tennessee Aquarium

1 There are 5 animals in the tank.
3 are clown fish.
How many are sea horses?

___5___ ⊖ ___3___ = ___2___

___2___ sea horses

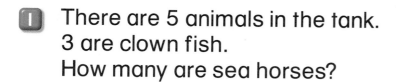

2 There are 8 animals in the tank.
4 are sea horses.
How many are clown fish?

_____ ◯ _____ = _____

_____ clown fish

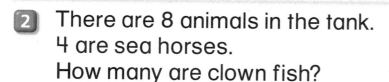

3 There are 9 animals in the tank.
8 are sea horses.
How many are not sea horses?

_____ ◯ _____ = _____

_____ not sea horses

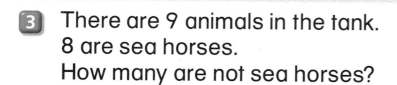

Unit 2 • Chapters 5–8

Name _____

CHALLENGE

Missing Parts

6 = __8__ − 2   6 = 8 − __2__

Add or subtract.
Write the missing numbers.

1.

   10 = ____ + 4   10 = 6 + ____

2.

   4 = ____ − 3   4 = 7 − ____

3.

   5 = ____ − 4   5 = 9 − ____

4.

   8 = ____ + 3   8 = 5 + ____

Name _____

✓ Study Guide and Review

Skills and Concepts

Use the number line. Circle the number you use to **count on**. Add.

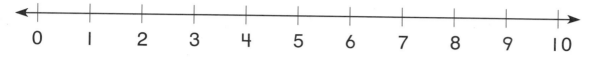

1. $6 + 2 =$ _____
2. $7 + 3 =$ _____

Use the number line. Circle the number you use to **count back**. Subtract.

3. $8 - 2 =$ _____
4. $9 - 3 =$ _____

Add. Then circle the doubles facts.

5. $2 + 2$
6. $8 + 2$
7. $2 + 5$
8. $4 + 4$
9. $3 + 3$
10. $7 + 2$

Add. Then subtract.

11. $5 + 4$ $9 - 4$
12. $7 + 3$ $10 - 7$
13. $4 + 4$ $8 - 4$

Unit 2 • Study Guide and Review

Add. Change the order. Write the new fact.

14. 4
    + 2
    ☐

    + ☐
    ☐

15. 1
    + 6
    ☐

    + ☐
    ☐

16. 3
    + 5
    ☐

    + ☐
    ☐

Add or subtract.
Write the numbers in the fact family.

17. 4 + 6 = ___   10 − 6 = ___

    6 + 4 = ___   10 − 4 = ___

Complete the table. Follow the rule.

18. | Add 3 | |
    |---|---|
    | 0 | |
    | 2 | |
    | 4 | |

19. | Subtract 2 | |
    |---|---|
    | 6 | |
    | 5 | |
    | 4 | |

20. | Subtract 1 | |
    |---|---|
    | 10 | |
    | 9 | |
    | 8 | |

Problem Solving

Circle **add** or **subtract**. Write the number sentence.

21. Max has 5 apples.
    Meg has 3 apples.
    How many do they have in all?

    ____ apples

    add   subtract

    ___ ○ ___ ○ ___

132 one hundred thirty-two

Name _____

Performance Assessment

Little Lambs

Maria went inside the barn.
Little lambs were in two stalls.

- She counted 3 little lambs in one stall.
- She counted some more little lambs in the other stall.
- Maria counted fewer than 7 little lambs in all.

Write a number sentence that fits this math story. Write the other number sentences in that fact family.

Show your work.

Name ______________________________

TECHNOLOGY

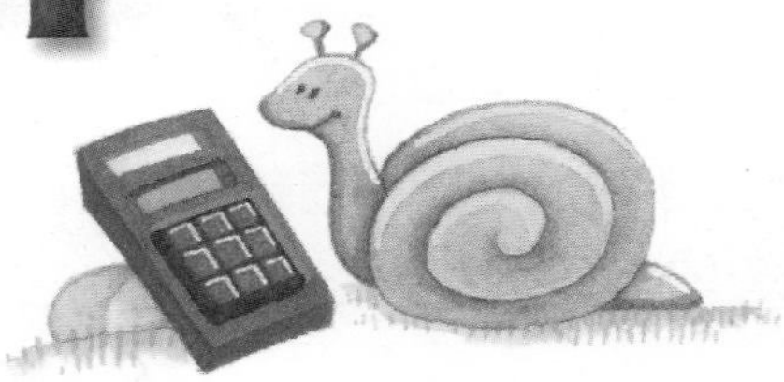

Calculator • Add and Subtract

Use a calculator.
Write the answers.

Press ON/C 2 + 3 = 5 − 1 = 4

Practice and Problem Solving

1. Press ON/C 6 + 4 = ______ − 1 = ______

2. Press ON/C 6 − 4 = ______ + 4 = ______

3. Press ON/C 3 + 0 = ______ + 7 = ______

4. Press ON/C 3 − 0 = ______ + 6 = ______

5. Press ON/C 7 + 2 = ______ + 1 = ______

6. Press ON/C 7 − 2 = ______ + 1 = ______

7. Press ON/C 6 − 3 = ______

Explain your answer.
Use ●.

TECHNOLOGY

SCHOOL HOME CONNECTION

Dear Family,

In Unit 2 we learned addition and subtraction facts to 10. Here is a game for us to play together. This game will give me a chance to share what I have learned.

Love,

_____

Directions
1. Put your game piece on START.
2. Use a paper clip and a pencil to make the spinner. Spin. Move that many spaces.
3. Add or subtract the number you spin and the number your game piece is on.
4. Move forward 1 if you added. Move forward 2 if you subtracted.
5. Take turns. The first person to get to END wins.

Materials
- 2 game pieces or beans
- pencil
- paper clip

Get Those Numbers

Spinner: 1, 2, 3, 4, 5

START 5 — 6 — 4 — 2 — 5 — 9 — 3 — 7 — Go again. 10

Go again. 3 — 4 — 5 — 3 — 2 — 5 — 7 — 2 — 4

6 — 3 — 1 — 5 — 4 — 8 — 6 — 6 — END

Unit 3 • Unit Game

one hundred thirty-five A 135A

SCHOOL HOME CONNECTION — LOOKING FORWARD

Dear Family,

During the next few weeks, we will learn about graphs and about numbers to 100. Here is important math vocabulary and a list of books to share.

Love,

_____

Vocabulary

picture graph
bar graph
is greater than
is less than
is equal to

Vocabulary Power

picture graph

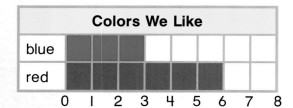

bar graph

Colors We Like								
blue								
red								

0 1 2 3 4 5 6 7 8

26 > 24
26 is greater than 24.

24 < 26
24 is less than 26.

24 = 24
24 is equal to 24.

BOOKS TO SHARE

To read about graphs and about numbers to 100 with your child, look for these books in your library.

Is It Rough? Is It Smooth? Is It Shiny?
by Tana Hoban,
Greenwillow, 1990.

Pancakes for Breakfast,
by Tomie dePaola,
Harcourt, 1990.

Two of Everything,
by Lily Toy Hong,
Albert Whitman, 1993.

Splash,
by Ann Jonas,
Greenwillow, 1995.

 Visit *The Learning Site* for additional ideas and activities. www.harcourtschool.com

CHAPTER 9: Graphs and Tables

FUN FACTS

You see ten fingers when you make handprint paintings.

Theme: Favorites

Name _____

✓ Check What You Know

Read a Tally Table

Fruits We Like	
🍌	|||||
🍎	|||
🍓	|||||	

Write how many.

1. _____

2. _____

3. _____

4. Circle the fruit the most children chose.

Make Picture Graphs

5. Look at the picture.
   Make a graph about pennies and nickels.

Pennies and Nickels

Write how many of each coin.
Circle the number that shows fewer.

6.   _____

7.    _____

136 one hundred thirty-six   Use this page to review important skills needed for this chapter.

Name _____

Algebra: Sort and Classify

Explore

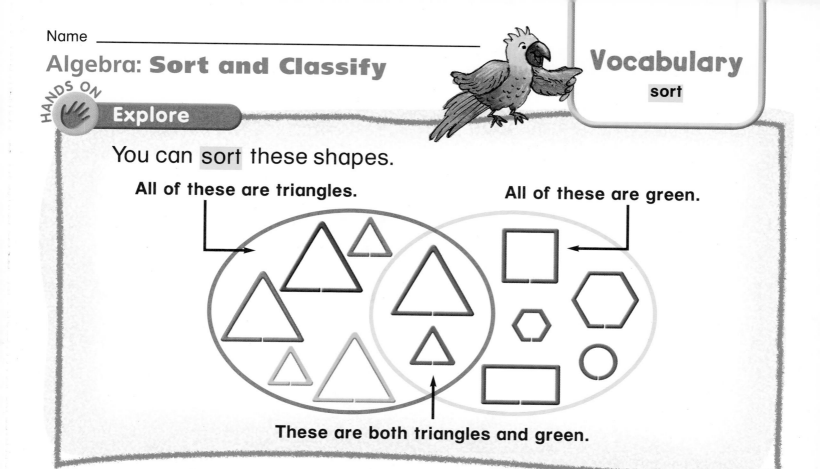

Vocabulary
sort

Connect

Sort your shapes a different way.
Draw each group. Tell how you sorted.

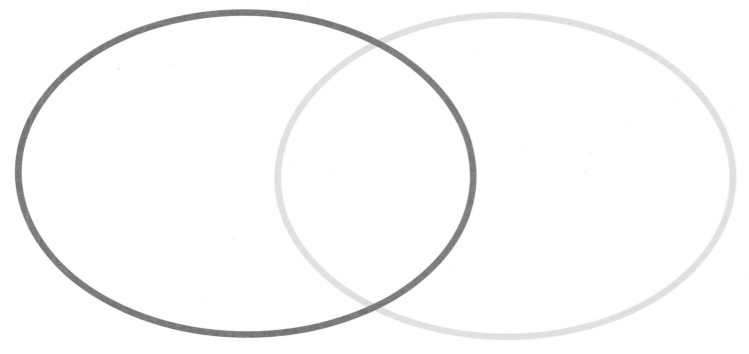

Explain It • Daily Reasoning

How could you sort your shapes into four different groups? Show one way.

Chapter 9 • Graphs and Tables

Practice and Problem Solving

Draw a line from each shape to the group where it belongs.

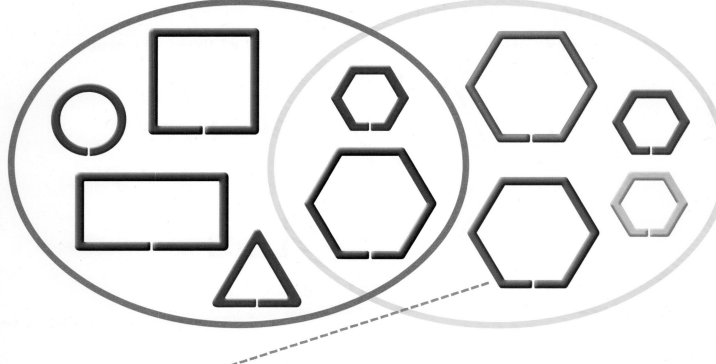

1.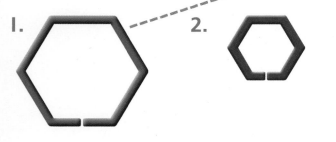

2.

3.

4.

Problem Solving
Application

5. Draw how you could sort these buttons.

Write About It • Draw how you could sort these buttons.

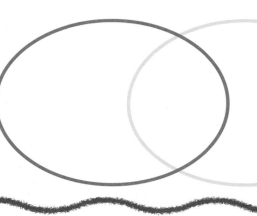

HOME ACTIVITY • Ask your child to explain how he or she sorted in Exercises 1–4.

Name _____

Make Concrete Graphs

Explore

Vocabulary
concrete graph

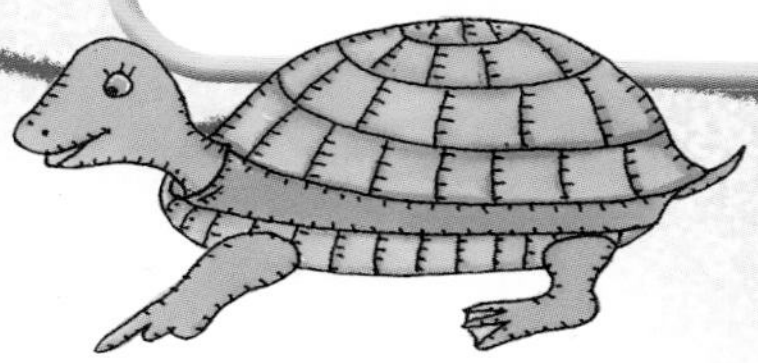

This concrete graph shows how many crayons there are of each color.

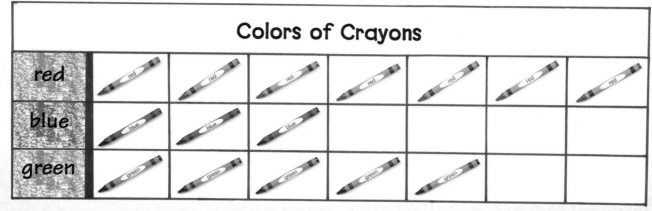

Connect

Sort , ▢, and ▢. Make a graph.

Colors of Cubes

1. How many ▢ are there? _____
2. How many ▢ are there? _____
3. How many ▢ are there? _____

Explain It • Daily Reasoning

Which color cube did you have the most of? How do you know?

Chapter 9 • Graphs and Tables

Name _____

Extra Practice

Children chose from three sports.

1. Write how many tally marks.

   Which sport do you like the best?

Sports We Like		Total
soccer	|||||	
baseball	||	
basketball	||||	

2. Color the bar graph to match the tally table.

Sports We Like					
soccer					
baseball					
basketball					

   0   1   2   3   4   5

Problem Solving

Use the bar graph to answer the questions.

3. How many children chose basketball?

   _____

4. Which game did the most children choose?

   _____

5. How many children chose soccer? _____

Chapter 9 • Graphs and Tables

one hundred fifty-one 151

Name _____

✓ Review/Test

Concepts and Skills

Children chose from three shirt colors.

1. Write how many tally marks.

   Which shirt color do you like the best?

Shirt Colors We Like		Total
red	|||||	
yellow	||||	
blue	|||||		

2. Color the bar graph to match the tally table.

Shirt Colors We Like						
red						
yellow						
blue						

   0 1 2 3 4 5 6

Problem Solving

Use the bar graph to answer the questions.

3. How many children chose red?

   _____

4. Which color did the most children choose?

   _____

5. How many children chose blue? _____

Name _____

★Standardized Test Prep
Chapters 1-9

Choose the answer for questions 1 – 3.
Use the graph to answer questions 1 and 2.

Our Pets					
dog	🐕	🐕	🐕	🐕	🐕
cat	🐈	🐈	🐈		
rabbit	🐇	🐇			

1. How many children have ?

   2   3   5   7
   ○   ○   ○   ○

2. Which pet do the fewest children have?

   ○   ○   ○

3. Which subtraction fact is related to 4 + 2 = 6?

   4 − 1 = 3   6 − 5 = 1   4 − 3 = 1   6 − 2 = 4
   ○   ○   ○   ○

Show What You Know

4. Sort. Fill in the tally table to explain how you sorted. Color the bar graph to match the tally marks.

Fruits We Like	Total
🍎	
🍌	
🍓	

Fruits We Like						
apple 🍎						
banana 🍌						
strawberry 🍓						
	0	1	2	3	4	5

5. How many more children chose apples than bananas?

   1   3   4   5
   ○   ○   ○   ○

Name _____

MATH GAME

Graph Game

Play with a partner.

1. Spin the 🕐.
2. Put 1 cube of that color in your graph.
3. Take turns until one player fills a row.
4. That player tosses the 🎲.
5. The player who has a row with that many cubes wins. If no player has a row with that many, toss again.

You will need

10 🎲
10 🎲
10 🎲

Player 1

Player 2

CHAPTER 10 Place Value to 100

SOCIAL STUDIES

FUN FACTS

A medium size piñata can hold about 100 pieces of hard candy and small toys.

Theme: Time for a Party

one hundred fifty-five 155

Name _____

✓ Check What You Know

Make Groups of 10

Count. Draw more to make a group of 10.

1.

2.

3.

4.

11 to 20: Using Ten Frames

Count. Circle the number that tells how many.
Write the number.

5. 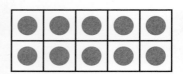   11   12   13   ____

6.    13   14   15   ____

7.     17   18   19   ____

8.     18   19   20   ____

156 one hundred fifty-six   Use this page to review important skills needed for this chapter.

Name _____

Teen Numbers

Vocabulary
ten
ones

Explore (Hands On)

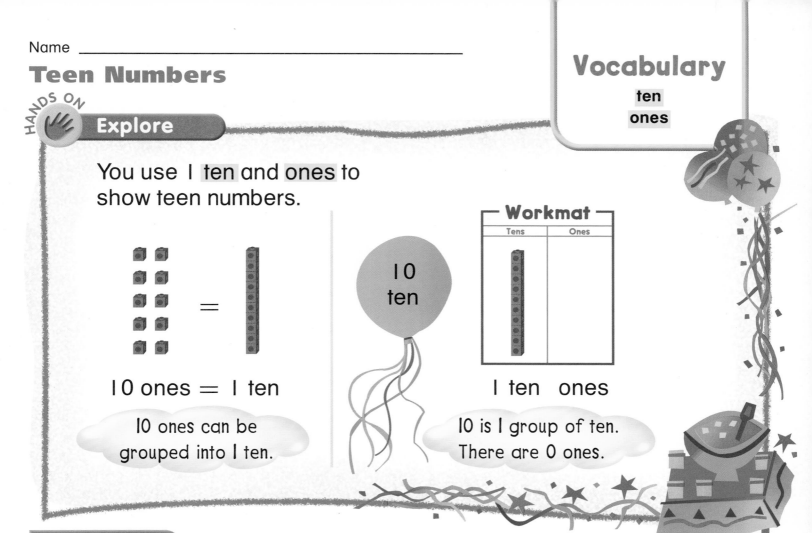

Connect

Use Workmat 3 and 🎲. Show the teen number. Draw the tens and ones. Write how many tens and ones.

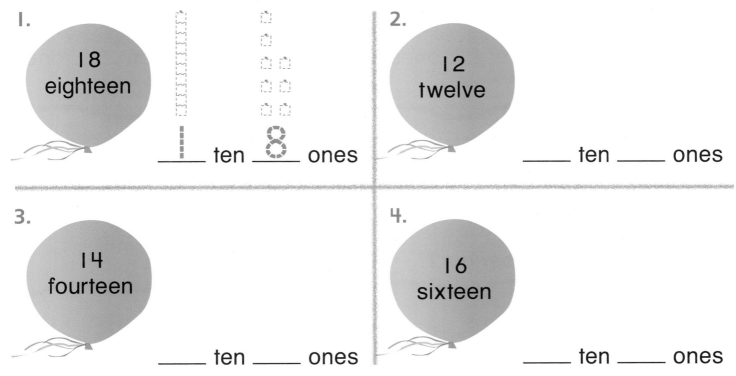

1. 18 eighteen
   __1__ ten __8__ ones

2. 12 twelve
   ____ ten ____ ones

3. 14 fourteen
   ____ ten ____ ones

4. 16 sixteen
   ____ ten ____ ones

Explain It • Daily Reasoning

What number comes before the first teen number?

Chapter 10 • Place Value to 100

Practice and Problem Solving

Draw the tens and ones.
Write how many tens and ones.

1. 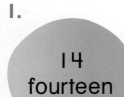 14 fourteen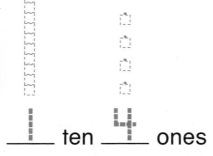
   __1__ ten __4__ ones

2. 19 nineteen
   ____ ten ____ ones

3. 13 thirteen
   ____ ten ____ ones

4. 11 eleven
   ____ ten ____ ones

5. 17 seventeen
   ____ ten ____ ones

6. 15 fifteen
   ____ ten ____ ones

Problem Solving

Logical Reasoning

Write the teen number that solves the riddle.

7. I am less than 19.
   I am more than 17.
   What teen number am I? _____

 Write About It • Look at Exercise 7.
Draw the tens and ones to show your answer.

⬠ **HOME ACTIVITY** • Have your child use small objects to show teen numbers between 10 and 20. Ask him or her to make groups of tens and ones, tell how many are in each group, and say the number.

Name _____

Extra Practice

Write how many tens. Write the number.

1.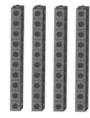

   ___ tens = ___

2.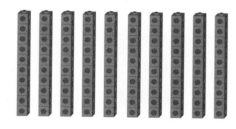

   ___ tens = ___

Write how many tens and ones. Write the number.

3.

   ___ ten ___ ones = ___

4.

   ___ tens ___ ones = ___

Write how many tens and ones. Write the number in a different way.

5.

   ___ tens ___ ones = ___

   ___ + ___

6.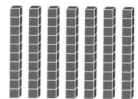

   ___ tens ___ ones = ___

   ___ + ___

Problem Solving

Circle the closest estimate.

7. About how many 🎲 would fill the bag?

   about 2 🎲

   about 20 🎲

   about 200 🎲

Chapter 10 • Place Value to 100

one hundred sixty-nine **169**

Name _____

✓ Review/Test

Concepts and Skills

Write how many tens. Write the number.

1.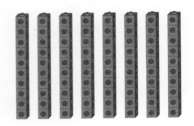

   ____ tens = ____

2.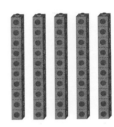

   ____ tens = ____

Write how many tens and ones. Write the number.

3.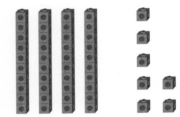

   ____ tens ____ ones = ____

4.

   ____ ten ____ ones = ____

Write how many tens and ones. Write the number in a different way.

5.

   ____ tens ____ one = ____

   ____ + ____

6.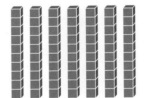

   ____ tens ____ ones = ____

   ____ + ____

Problem Solving

Circle the closest estimate.

7. About how many 🟦 would fill the cup?

   about 1 🟦

   about 10 🟦

   about 100 🟦

170 one hundred seventy

Name _____

★Standardized Test Prep
Chapters 1–10

Choose the answer for questions 1–4.

1. Which is another way to write the number?

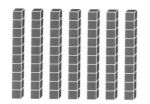

7 tens + 6 ones = 76

- ○ 60 + 7
- ○ 70 + 3
- ○ 70 + 2
- ○ 70 + 6

2. Which does the picture show?

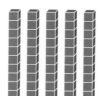

- ○ 3 tens
- ○ 5 tens
- ○ 7 tens
- ○ 9 tens

3. Which is the closest estimate? About how many are in your classroom?

- ○ about 3
- ○ about 30
- ○ about 300

4. Which fact is related to 5 − 2 = 3?

- ○ 1 + 4 = 5
- ○ 5 + 3 = 8
- ○ 3 + 2 = 5
- ○ 3 − 2 = 1

Show What You Know

5. Choose a number. Draw to explain how many tens and ones. Write the number two ways.

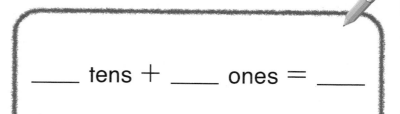

____ tens + ____ ones = ____

____ + ____ = ____

_____

Chapter 10

MATH GAME

Teen Spin

Play with a partner.

1. Put all the 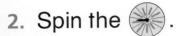 in the penny pile.
2. Spin the .
3. Read the teen number word.
4. Tell how many more than 10 that number is.
5. Put that many  in your bank.
6. Take turns until the penny pile is gone.
7. The player with more 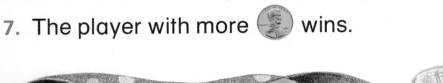 wins.

You will need

Penny Pile

CHAPTER 11
Comparing and Ordering Numbers

FUN FACTS

The first hopscotch court was 100 feet long.

Theme: Outdoor Fun

Name _____

✓ Check What You Know

More, Less, Same
Count. Write the number.

1. Circle the number that is less.

_____   _____

2. Circle the number that is more.

_____   _____

3. Circle the 2 numbers that are the same.

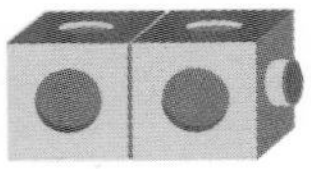

_____   _____   _____

Order Numbers on a Number Line
Write the missing numbers.

4.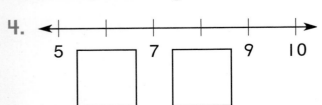

5.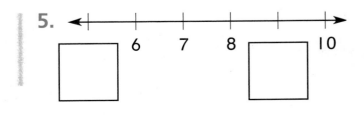

Name _____

Algebra: Greater Than

Vocabulary
is greater than >

Explore

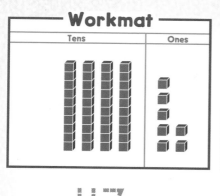
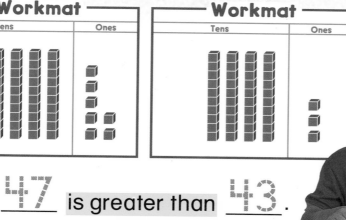

43 and 47 have the same number of tens, but 47 has more ones.

__47__ is greater than __43__.

__47__ > __43__

Connect

Use ▭▭▭ ▫ to show each number.
Circle the greater number. Write the numbers.

1.

   __53__ is greater than __35__.

   __53__ > __35__

2.

   ____ is greater than ____.

   ____ > ____

3.

   ____ is greater than ____.

   ____ > ____

4.

   ____ is greater than ____.

   ____ > ____

Explain It • Daily Reasoning

To find the greater number, should you look first at the ones or at the tens? Why?

Practice and Problem Solving

Circle the greater number.
Write the numbers.
You can use ▭▭▭▭▭ ▪ .

THINK:
Look at the tens place first.

1.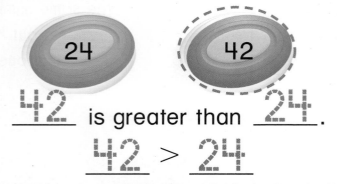

   _42_ is greater than _24_.

   _42_ > _24_

2.  76   78

   _____ is greater than _____.

   _____ > _____

3.  51   15

   _____ is greater than _____.

   _____ > _____

4.  69   98

   _____ is greater than _____.

   _____ > _____

5.  37   35

   _____ is greater than _____.

   _____ > _____

6.  54   45

   _____ is greater than _____.

   _____ > _____

Problem Solving
Application

7. Circle the numbers that are greater than 50.

   14   83   94   44   62   70

 Write About It • Use one of the numbers you circled in Exercise 7 to complete ☐ > 50.

🏠 **HOME ACTIVITY** • Say two numbers that are less than 100. Ask your child which is greater. Then have him or her write the two numbers, using the *greater than* symbol (>).

Name _____

Extra Practice

Circle the greater number.
Write the numbers.

1.   ____ > ____

Circle the number that is less.
Write the numbers.

2.   ____ < ____

3. Count forward.   68, ____, ____

4. Count backward.   41, ____, ____

Write the number that is just before, between,
or just after.

5. 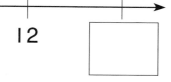   6.

7.   8.   9.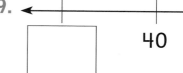

Problem Solving

Use the model 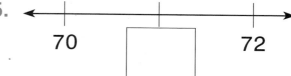. Find 10
more or 10 less. Write the number.

10. Brad has 23 toy cars. Ken
has 10 less. How many toy
cars does Ken have?

____ toy cars

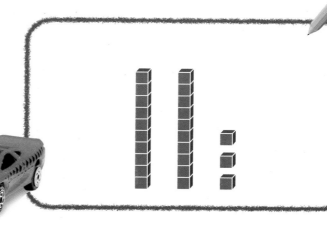

Chapter 11 • Comparing and Ordering Numbers   one hundred eighty-seven **187**

✓ Review/Test

Concepts and Skills

Circle the greater number.
Write the numbers.

1.   _____ > _____

Circle the number that is less.
Write the numbers.

2.   _____ < _____

3. Count forward.   90, _____, _____

4. Count backward. 63, _____, _____

Write the number that is just before, between, or just after.

5.

6.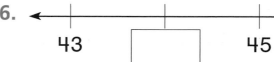

7. 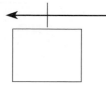   8. 59 _____   9.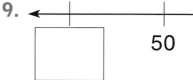

Problem Solving

Use the model ▬▬ ▪. Find 10 more or 10 less. Write the number.

10. Jose has 65 jacks. Elsa has 10 more. How many jacks does Elsa have?

_____ jacks

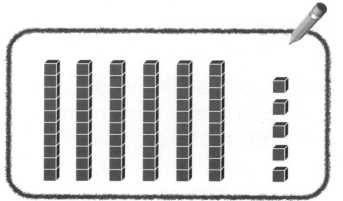

Name _____

★Standardized Test Prep
Chapters 1-11

Choose the answer for questions 1–4.

1. Which number is between 33 and 35?

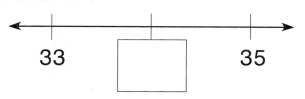

  32   34   36   37
  ○   ○   ○   ○

2. Which number is less than 24?

  19   27   30   41
  ○   ○   ○   ○

3. Which numbers are missing?

  __20__, __19__, _____, _____, _____

  18, 17, 16   19, 20, 21   21, 22, 23   21, 20, 19
  ○   ○   ○   ○

4. Which is another way to write 9 − 3 = 6?

  $\begin{array}{r}9\\-4\\\hline 5\end{array}$ $\begin{array}{r}9\\-3\\\hline 6\end{array}$ $\begin{array}{r}9\\-2\\\hline 7\end{array}$ $\begin{array}{r}9\\-1\\\hline 8\end{array}$
  ○   ○   ○   ○

Show What You Know

5. Find 10 less. Use the model to explain your answer.

   April has 37 marbles.
   Molly has 10 less.
   How many marbles does Molly have?

   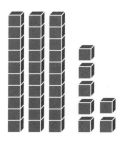

   _____ marbles

Chapter 11   one hundred eighty-nine **189**

MATH GAME

Greater Steps

Play with a partner.

1. Put your 🎮 at START.
2. Toss the 🎲 and the 🎲.
3. Use one number as tens and one number as ones. Can you make a number that solves the problem on your next step?
4. If you can, move your 🎮 up to that step.
5. If you can not, your turn is over.
6. The first player to get to END wins.

You will need

2 🎮

2 🎲

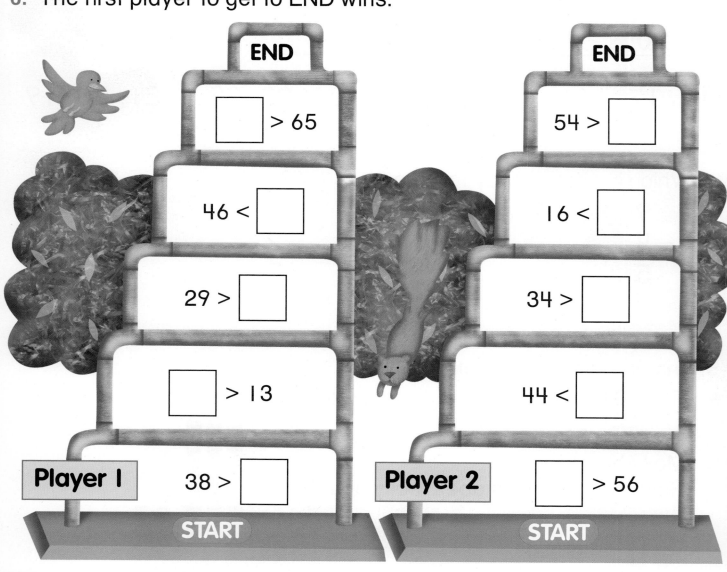

Player 1: 38 > ☐, ☐ > 13, 29 > ☐, 46 < ☐, ☐ > 65, END

Player 2: ☐ > 56, 44 < ☐, 34 > ☐, 16 < ☐, 54 > ☐, END

CHAPTER 12 Number Patterns

FUN FACTS

The number of crayons in some boxes are all even numbers: 8, 16, 24, 32, 48, 64, and 96.

Theme: Art Class

Name _____

✓ Check What You Know

Count Orally Using a Hundred Chart

Touch and count. Color the last number counted.

1. Start at 1 and count to 30.
2. Start at 40 and count to 53.
3. Start at 92 and count to 98.

1	2	3	4	5	6	7	8	9	10
11	12	13	14	15	16	17	18	19	20
21	22	23	24	25	26	27	28	29	30
31	32	33	34	35	36	37	38	39	40
41	42	43	44	45	46	47	48	49	50
51	52	53	54	55	56	57	58	59	60
61	62	63	64	65	66	67	68	69	70
71	72	73	74	75	76	77	78	79	80
81	82	83	84	85	86	87	88	89	90
91	92	93	94	95	96	97	98	99	100

Ordinal Numbers

4. Circle the second cat.
5. Mark an X on the fifth cat.
6. Draw a line under the first cat.

one hundred ninety-two   Use this page to review important skills needed for this chapter.

Name _____

Skip Count by 2s, 5s, and 10s

Learn

Skip count. Count the jars of paint by twos. Write how many.

2   4   6   ____   ____

12   ____   ____   ____   ____ jars of paint

Check

Skip count. Count the fingers by fives. Write how many.

1.

5   10   ____   ____   ____   ____

35   ____   ____   ____   ____   ____ fingers

Skip count. Count the toes by tens. Write how many.

2.

10   20   ____   ____   ____

____   ____   ____   ____   ____ toes

Explain It • Daily Reasoning

What number comes after 100 when you skip count by tens? Explain.

Chapter 12 • Number Patterns

Practice and Problem Solving

Skip count. Write how many.

1.

   2 ____ ____ ____ ____ ____ ____

2.

   5 ____ ____ ____ ____ ____ ____

3.

   10 ____ ____ ____ ____ ____ ____

Skip count. Write the missing numbers.

4. 2, 4, 6, ____, ____, ____, 14, ____, 18

5. 10, ____, 30, ____, ____, 60, 70, ____, 90

Problem Solving
Visual Thinking

6. Skip count.
   Each hand has 5 fingers.
   How many fingers are there in all?

   ____ fingers

 Write About It • Draw a picture to show skip counting by twos. Write the numbers.

> **HOME ACTIVITY** • Draw 20 stars. Have your child circle groups of 2 and then count by twos to find the total. Repeat the activity for groups of 5.

Name _____

Algebra: Use a Hundred Chart to Skip Count

Learn

Write the missing numbers. Count by tens.
Use 🖍 to color the numbers you say.

Start on 10. Count forward by tens.

1	2	3	4	5	6	7	8	9	10
11	12	13	14	15	16	17	18	19	
21	22	23	24	25	26	27	28	29	
31	32	33	34	35	36	37	38	39	
41	42	43	44	45	46	47	48	49	
51	52	53	54	55	56	57	58	59	
61	62	63	64	65	66	67	68	69	
71	72	73	74	75	76	77	78	79	
81	82	83	84	85	86	87	88	89	
91	92	93	94	95	96	97	98	99	

Check

1. Count again. Count by fives.
   Use 🖍 to color the numbers you say.

Explain It • Daily Reasoning

What patterns do you see?
How could you make a new pattern?

Chapter 12 • Number Patterns   one hundred ninety-five **195**

Name _____

Standardized Test Prep
Chapters 1–12

1. Count by fives. Which number is missing?

5, 10, 15, 20, ___, 30

25 35 40 45
○   ○   ○   ○

2. Count by twos. Which number is missing?

2, 4, ___, 8, 10

3   5   6   12
○   ○   ○   ○

3. Which number is even?

3   5   6   7
○   ○   ○   ○

4. Which is fourth?

first

○ ○ ○ ○ ○

5. Which number is just before 28?

27   28   29   36
○   ○   ○   ○

Show What You Know

6. How many ears are on 3 cats? Draw a picture to explain your answer.

Number of Cats	1	2	3
Number of Ears	2	4	

Chapter 12   two hundred seven **207**

MATH GAME

Even Skips

Play with a partner.

1. Put your ♟ on START.
2. Toss the 🎲.
3. Move your ♟ that many spaces.
4. If you land on an odd number, your turn is over.
5. If you land on an even number, skip-count on by twos 3 times. Move to that number.
6. The first player to get to END wins.

You will need

🎲   2 ♟

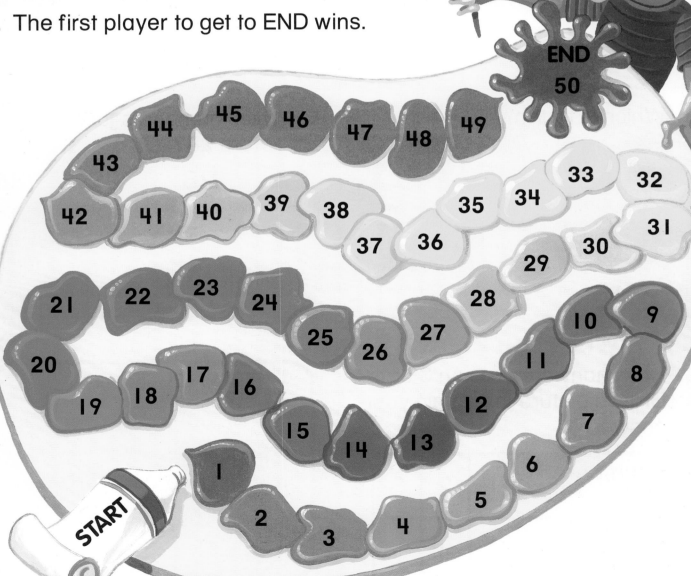

CHAPTER 13
Addition and Subtraction Facts to 12

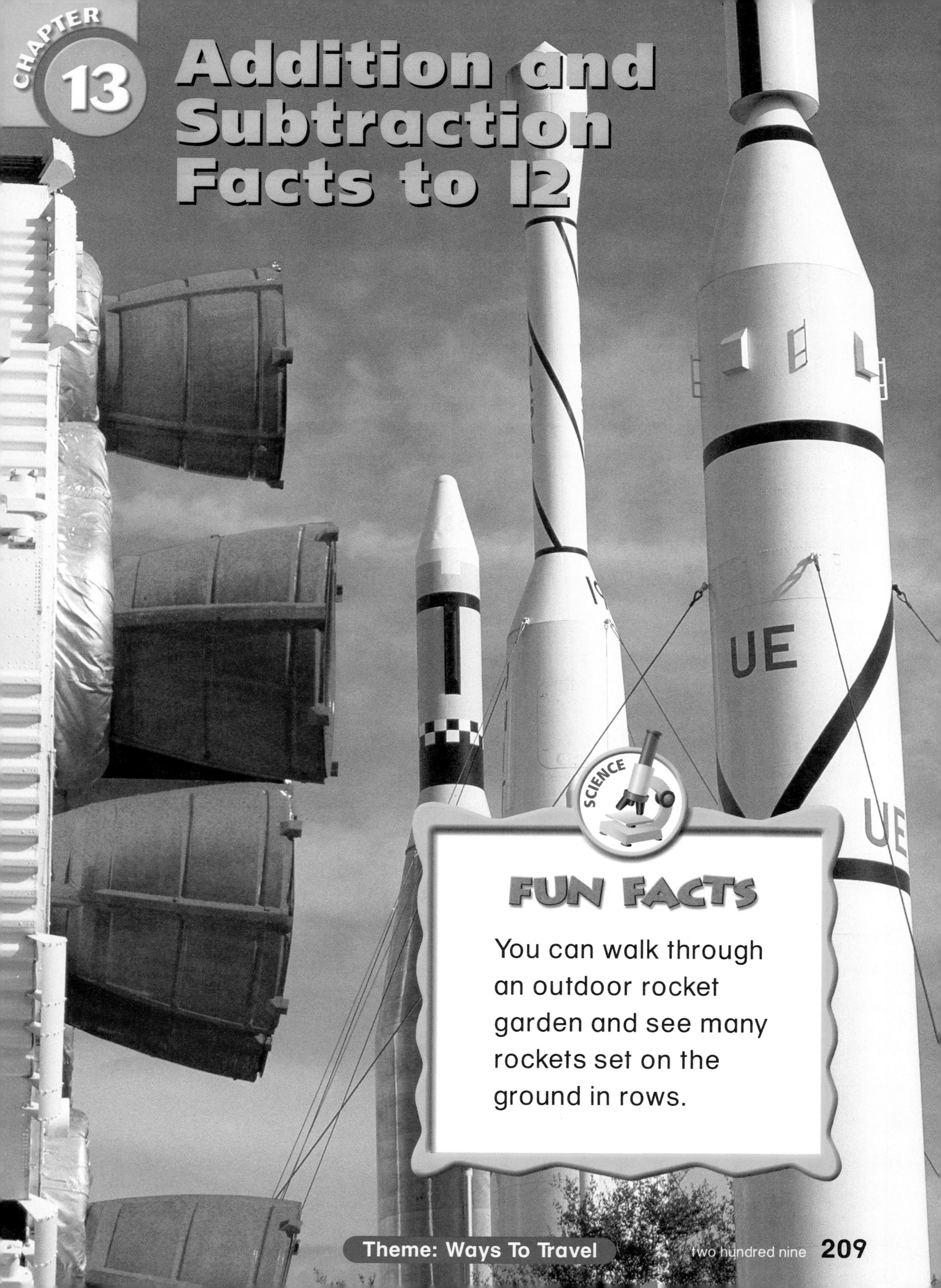

FUN FACTS

You can walk through an outdoor rocket garden and see many rockets set on the ground in rows.

Theme: Ways To Travel

Name _____

✓ Check What You Know

Use a Number Line to Count On

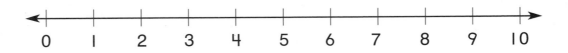

Use the number line. Count on to find the sum.

1. 2
  +2

2. 4
  +1

3. 6
  +3

4. 5
  +2

5. 8
  +1

6. 7
  +3

Use Doubles

Add. Then circle the doubles facts.

7. 4
  +3

8. 4
  +4

9. 2
  +3

10. 5
  +5

11. 2
  +4

12. 3
  +3

Use a Number Line to Count Back

Use the number line. Count back to subtract.

13. 9
  −3

14. 6
  −3

15. 10
  − 2

16. 8
  −3

17. 4
  −2

18. 10
  − 1

210 two hundred ten   Use this page to review important skills needed for this chapter.

Name _____

Count On to Add

Vocabulary
count on

Learn

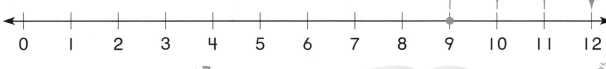

9
+3

12

Start on 9. Then move 3 spaces to the right.
10, 11, 12

Start with the greater number.
Count on to add.

Check

Circle the greater number.
Use the number line. Count on to add.

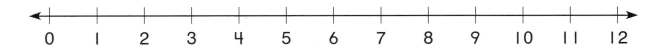

1. ⑤
   +3
   ---
2. 2
   +7
   ---
3. 3
   +8
   ---
4. 9
   +1
   ---
5. 3
   +9
   ---
6. 7
   +3
   ---

7. 8
   +1
   ---
8. 2
   +9
   ---
9. 9
   +2
   ---
10. 8
    +2
    ---
11. 1
    +7
    ---
12. 4
    +3
    ---

13. 3
    +7
    ---
14. 5
    +2
    ---
15. 3
    +6
    ---
16. 8
    +3
    ---
17. 6
    +2
    ---
18. 2
    +8

Explain It • Daily Reasoning

Why is it easier to start with the greater number when you count on?

Chapter 13 • Addition and Subtraction Facts to 12

Practice and Problem Solving

$3 + \text{⑧} = \underline{11}$

Start on 8. Move 3 spaces to the right.
9, 10, 11

Circle the greater number.
Use the number line. Count on to add.

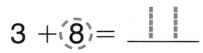

1. ⑥ + 2 = ___
2. 3 + 7 = ___
3. 9 + 3 = ___
4. 1 + 8 = ___
5. 6 + 3 = ___
6. 8 + 2 = ___
7. 2 + 9 = ___
8. 3 + 5 = ___
9. 7 + 2 = ___
10. 3 + 4 = ___
11. 9 + 1 = ___
12. 5 + 2 = ___

Problem Solving
Application

Count on to solve. Draw a picture to check.

13. Ann Lee saved 6¢.
Carol saved 3¢.
How much did
they save in all? ___ ¢

 Write About It • Make up an addition story about this picture. Then write the number sentence.

HOME ACTIVITY • With your child, make flash cards for the addition facts for sums through 12. Practice the facts together.

212 two hundred twelve

Name _____

Doubles and Doubles Plus 1

Vocabulary
doubles
doubles plus one

Learn

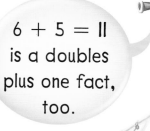

"6 + 5 = 11 is a doubles plus one fact, too."

```
  5          5
 +5         +6
 ──         ──
 10         11
```

5 + 5 = 10 is a **doubles** fact.

5 + 6 = 11 is a **doubles plus one** fact.

Check

Write the three sums.
Then circle the doubles fact.

1. (3+3=6)   3+4   4+3   2. 4+4   4+5   5+4

3. 2+2   2+3   3+2   4. 1+1   1+2   2+1

5. 0+0   0+1   1+0   6. 5+5   5+6   6+5

Explain It • Daily Reasoning

How does knowing the sum for 4 + 4 help you find the sums for 4 + 5 and 5 + 4?

Chapter 13 • Addition and Subtraction Facts to 12   two hundred thirteen **213**

Practice and Problem Solving

Write the sums.

1. 2 + 2 = __4__,   so 2 + 3 = __5__

2. 1 + 1 = _____,   so 2 + 1 = _____

3. 5 + 5 = _____,   so 5 + 6 = _____

4. 4 + 4 = _____,   so 5 + 4 = _____

5. 0 + 0 = _____,   so 0 + 1 = _____

6. 3 + 3 = _____,   so 3 + 4 = _____

Problem Solving

Logical Reasoning

Solve. Draw a picture to check.

7. Pat has 3 .
   Bob has double that many.
   Sue has double what Bob has.

   How many does each person have?

   _____   _____   _____

   Pat   Bob   Sue

Write About It • Look at Exercise 7.
Explain how you got your answers.

 HOME ACTIVITY • Have your child tell you the doubles facts and the doubles plus one facts for 2, 3, 4, and 5. For example, 2 + 2 = 4, so 2 + 3 = 5.

Name _____

Algebra: Add 3 Numbers

Explore

4
2 + 2 + 6 = 10

8
2 + 2 + 6 = 10

The sums are the same!

Connect

Use . Add the blue numbers first.
Write the sums.

1. 3 + 2 + 5 = ____   3 + 2 + 5 = ____

2. 5 + 2 + 2 = ____   5 + 2 + 2 = ____

3. 9 + 0 + 3 = ____   9 + 0 + 3 = ____

4. 3 + 3 + 2 = ____   3 + 3 + 2 = ____

5. 1 + 5 + 3 = ____   1 + 5 + 3 = ____

6. 3 + 6 + 2 = ____   3 + 6 + 2 = ____

Explain It • Daily Reasoning

The numbers to be added are 6, 2, and 1.
Which numbers will you add first? Why?

Chapter 13 • Addition and Subtraction Facts to 12

Practice and Problem Solving

Circle the two numbers you add first. Write the sum.

1. (2) 3 +4 = 9 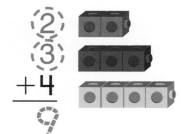   5 +4 = 9
2. 2 (3) (4) = 9 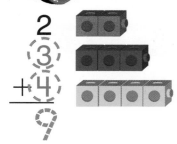   2 +7 = 9

3. 2
    5
   +1

4. 3
    4
   +3

5. 1
    1
   +6

6. 7
    2
   +1

7. 2
    1
   +6

8. 6
    1
   +4

9. 3
    5
   +2

10. 2
    5
   +5

11. 3
    2
   +3

12. 4
    5
   +0

Problem Solving
Mental Math

Add in your head. Draw to check.

13. Tim has 2 dogs. Steve has 3 cats. Greg has 6 fish. How many pets do the boys have in all?

    _____ pets

 Write About It • Look at Exercise 13. Which numbers did you decide to add first? Tell why.

HOME ACTIVITY • Have your child use small objects to show how he or she found the sum for each exercise on this page.

Problem Solving Strategy
Write a Number Sentence

<u>5 children ride bicycles.</u>

<u>4 children join them.</u>

(How many children are riding bicycles now?)

UNDERSTAND

What do you want to find out?

Circle the question.

PLAN

What facts do you need?

Underline them.

SOLVE

Write a number sentence to solve.

    children

(5 + 4 = 9)

CHECK

Does your answer make sense?

Draw a picture to check.

Write a number sentence.
Draw a picture to check.

THINK: Where do I put the numbers in my sentence?

1. Jim saw 4 boats. Then he saw 3 more. How many boats did he see in all?

 boats

Problem Solving Practice

Write a number sentence.
Draw a picture to check.

Keep in Mind!
Understand
Plan
Solve
Check

THINK: Where do I put the numbers in my sentence?

1. 5 children run.
   6 more join them.
   How many children are running now?

   ___◯___◯___ children

2. Lilly sees 7 cars.
   Then she sees 3 more.
   How many cars does she see in all?

   ___◯___◯___ cars

3. 9 children go for a walk.
   3 more children join them.
   How many children are walking now?

   ___◯___◯___ children

4. Ross has 2 toy rockets.
   Ida gives him 7 more.
   How many toy rockets does Ross have now?

   ___◯___◯___ toy rockets

HOME ACTIVITY • Give your child 1 to 9 small objects all alike, such as paper clips. Then add several more, to a total of no more than 12. Have your child write an addition sentence about them. Repeat, using different numbers.

Name _____

Count Back to Subtract

Vocabulary
count back

Learn

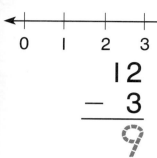

```
  12
-  3
----
   9
```

Start at 12. Then move 3 spaces to the left.
11, 10, 9

Start at 12. Count back to subtract.

Check

Use the number line to count back.
Write the difference.

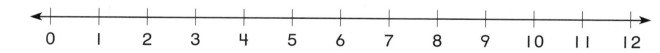

1. 10 − 2
2. 8 − 1
3. 6 − 2
4. 9 − 2
5. 7 − 2
6. 8 − 2

7. 11 − 3
8. 10 − 3
9. 9 − 3
10. 6 − 1
11. 8 − 3
12. 7 − 1

13. 10 − 1
14. 11 − 2
15. 9 − 1
16. 7 − 3
17. 6 − 3
18. 5 − 3

Explain It • Daily Reasoning

How can you find the difference for 10 − 3 without using a number line?

10 − 3

Chapter 13 • Addition and Subtraction Facts to 12

two hundred nineteen **219**

Practice and Problem Solving

Count back to subtract.

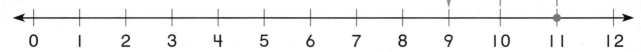

$11 - 2 = \underline{9}$

Start at 11.
Then move 2 spaces
to the left.
10, 9

Count back to subtract. Write the difference.
You can use the number line to help.

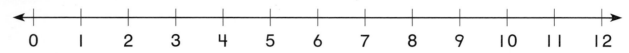

1. $12 - 3 = \underline{}$
2. $9 - 3 = \underline{}$
3. $10 - 2 = \underline{}$
4. $9 - 2 = \underline{}$
5. $10 - 3 = \underline{}$
6. $11 - 3 = \underline{}$
7. $10 - 1 = \underline{}$
8. $11 - 2 = \underline{}$
9. $6 - 3 = \underline{}$
10. $7 - 3 = \underline{}$
11. $9 - 1 = \underline{}$
12. $8 - 3 = \underline{}$

Problem Solving
Application

Write a number sentence to solve.
Use the number line to help.

13. There are 12 cars in the parking lot. 2 cars leave. How many cars are still in the parking lot? ____ ◯ ____ ◯ ____ cars

Write About It • Look at Exercise 13. Explain how you used the number line to subtract.

HOME ACTIVITY • Help your child use the number line on this page to practice any subtraction facts he or she missed in this lesson.

Name _____

Subtract to Compare

Learn

How many more red cars than blue cars are there?

There are 2 more red cars than blue cars.

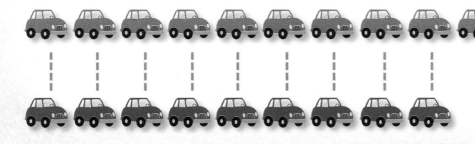

$$\begin{array}{r} 11 \\ -9 \\ \hline 2 \end{array}$$

Check

Draw lines to match. Write the difference.

1. How many fewer small boats than big boats are there?

$$\begin{array}{r} 12 \\ -8 \\ \hline \end{array}$$

2. How many more green bikes than yellow bikes are there?

$$\begin{array}{r} 10 \\ -7 \\ \hline \end{array}$$

Explain It • Daily Reasoning

Which group in the picture has more? How can you prove your answer?

Chapter 13 • Addition and Subtraction Facts to 12   two hundred twenty-one **221**

Practice and Problem Solving

Draw lines to match.
Write the difference.

1. 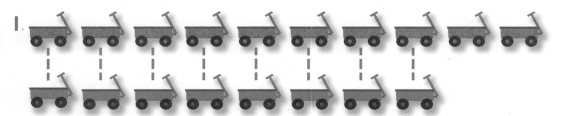   10
   − 8

   2

2.   11
   − 7

3.   12
   − 9

4.   12
   − 7

Problem Solving
Application

Solve. Draw a picture to check.

5. Margie has 10 cars.
   Jake has 6 cars.
   How many more cars
   does Margie have?

   _____ more cars

Write About It • Look at Exercise 5.
Explain how you got your answer.

HOME ACTIVITY • Set out two groups of objects, one with more objects than the other. Have your child show how to use matching to subtract to find out how many more are in the larger group.

Name _____

Extra Practice

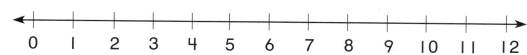

Circle the greater number.
Count on to add.

1. 8
  +3

2. 8
  +2

3. 2
  +5

Count back to subtract.
Write the difference.

4. 9
  −2

5. 8
  −3

6. 10
  − 3

Write all the sums. Then circle the doubles facts.

7. 2 + 2 = ___

8. 4 + 3 = ___

9. 5 + 5 = ___

Circle the two numbers you added first.
Write the sum.

10. 3 + 2 + 1 = ___

11. 7 + 1 + 2 = ___

Draw lines to match.
Write the difference.

12.

10
− 6

Problem Solving

Write a number sentence.
Draw a picture to check.

13. 6 children are on the bus.
    2 more children get on.
    How many children are
    on the bus?

 children

Chapter 13 • Addition and Subtraction Facts to 12

two hundred twenty-three **223**

Name _____

✓ Review/Test

Concepts and Skills

Circle the greater number.
Count on to add.

Count back to subtract.
Write the difference.

1. 9
   +3

2. 9
   +2

3. 3
   +6

4. 11
   − 2

5. 10
   − 3

6. 12
   − 3

Write all the sums. Then circle the doubles facts.

7. 4 + 4 = ___

8. 5 + 4 = ___

9. 3 + 3 = ___

Circle the two numbers you add first.
Write the sum.

10. 5 + 5 + 2 = ____

11. 4 + 4 + 2 = ____

Draw lines to match.
Write the difference.

12.

10
− 4

Problem Solving

Write a number sentence.
Draw a picture to check.

13. 7 children run. 3 more children join them. How many children are running now?

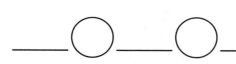 children

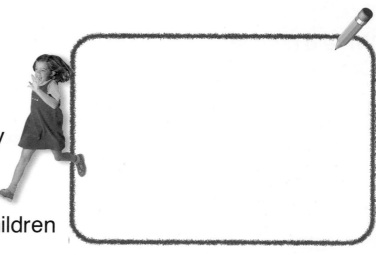

224 two hundred twenty-four

Name _____

★ Standardized Test Prep
Chapters 1–13

1. $\begin{array}{r} 10 \\ -3 \\ \hline \end{array}$ 5   7   13   15
                   ○   ○   ○   ○

2. Which does the number line show?

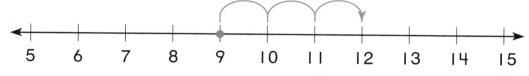

   $9 - 2 = 7$ $9 - 3 = 6$ $2 + 9 = 11$ $3 + 9 = 12$
      ○   ○   ○   ○

3. Which is a doubles plus 1 fact?

   $3 + 3 = 6$ $3 + 4 = 7$ $5 + 5 = 10$ $7 + 4 = 11$
      ○   ○   ○   ○

4. Which is the sum?
                     8   9   10   11
   $3 + 4 + 2 =$ ___   ○   ○   ○   ○

5. Which tells how many more apples than pears there are?

                          $7 - 5 = 2$ $8 - 5 = 3$
                              ○   ○

                          $5 + 2 = 7$ $5 + 3 = 8$
                              ○   ○

Show What You Know

6. Write the number sentence. Draw a picture to explain your answer.

   Kate picks 4 pumpkins. Edna picks 5 pumpkins. How many pumpkins do they pick in all?

   ___ ○ ___ ○ ___ pumpkins

Chapter 13   two hundred twenty-five **225**

MATH GAME

Boats Full of Facts

Play with a partner.

1. Put your 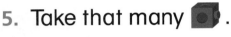 at START.
2. Spin the .
3. Move your that many boats.
4. Find the sum or difference.
5. Take that many .
6. When the first player gets to END, count .
7. The player with more wins.

You will need

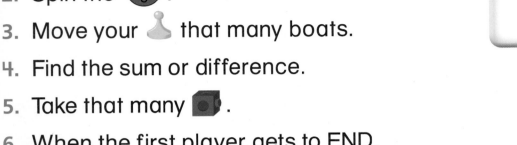

CHAPTER 14
Practice Addition and Subtraction

FUN FACTS

A square picnic table with 4 long benches can seat between 10 to 12 people.

Theme: A Picnic

Name _____

✓ Check What You Know

Sums to 10

Add. Write the sums.

1. 2   8
   +8   +2

2. 6   3
   +3   +6

3. 7   1
   +1   +7

4. 4   5
   +5   +4

5. 3   7
   +7   +3

6. 8   0
   +0   +8

Subtraction to 10

Subtract. Circle the pair of facts if they use the same numbers.

7. 9   9
   −2   −7

8. 8   8
   −0   −8

9. 10   10
   − 4   − 5

10. 7   7
   −6   −2

11. 8   8
   −2   −6

12. 10   10
   − 3   − 7

Subtract across. Subtract down.

13.

9	4	
6	2	

14.

10	3	
8	2	

228 two hundred twenty-eight   Use this page to review important skills needed for this chapter.

Name _____

Algebra: Related Addition and Subtraction Facts

Vocabulary
related facts

Explore

Related facts use the same numbers.

$8 + 4 = 12$

$12 - 4 = 8$

Connect

Use 🟥 🟧 to show related facts.
Complete the chart.

	Use 🟥	Add 🟧	Write the sum.	Take away	Write the subtraction sentence.
1.	6	3	$6 + 3 = \underline{9}$	3	$\underline{9} \bigcirc \underline{3} \bigcirc \underline{6}$
2.	7	5	$7 + 5 = \underline{}$	7	$\underline{} \bigcirc \underline{} \bigcirc \underline{}$
3.	4	6	$4 + 6 = \underline{}$	4	$\underline{} \bigcirc \underline{} \bigcirc \underline{}$
4.	3	8	$3 + 8 = \underline{}$	8	$\underline{} \bigcirc \underline{} \bigcirc \underline{}$
5.	6	6	$6 + 6 = \underline{}$	6	$\underline{} \bigcirc \underline{} \bigcirc \underline{}$

Explain It • Daily Reasoning

How can knowing an addition fact help you remember a subtraction fact?

Chapter 14 • Practice Addition and Subtraction

Practice and Problem Solving

Write each sum or difference.
Circle the related facts in each row.

1. (7 + 2 = 9)   5 + 2 = 7   (9 − 2 = 7)

2. 8 + 4 = ___   12 − 4 = ___   10 − 4 = ___

3. 11 − 4 = ___   9 + 1 = ___   10 − 1 = ___

4. 10 − 7 = ___   3 + 7 = ___   7 + 4 = ___

5. 6 + 5 = ___   12 − 5 = ___   11 − 5 = ___

6. 9 + 3 = ___   8 − 4 = ___   12 − 3 = ___

Problem Solving
Algebra

7. Circle three numbers that you can use to write a pair of related facts. Write the number sentences.

 5   7   8   12

___ ○ ___ ○ ___ | ___ ○ ___ ○ ___

 Write About It • Look at Exercise 7. Write the other related addition and subtraction sentences.

 HOME ACTIVITY • Give your child an addition problem, such as 4 + 5, and ask him or her to tell you the sum (9). Then ask your child to tell you a related subtraction fact (9 − 5 = 4 or 9 − 4 = 5).

Name _____

Fact Families to 12

Vocabulary
fact family

Explore

8 + 3 = 11

3 + 8 = 11

8, 3, and 11 are the numbers in this fact family.

11 − 3 = 8

11 − 8 = 3

Connect

Use 🎲 🎲. Add or subtract.
Write the numbers in the fact family.

1. 7 + 4 = __11__
    4 + 7 = __11__
    11 − 4 = __7__
    11 − 7 = __4__
    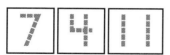
    | 7 | 4 | 11 |

2. 8 + 4 = ___
    4 + 8 = ___
    12 − 4 = ___
    12 − 8 = ___

3. 5 + 4 = ___
    4 + 5 = ___
    9 − 4 = ___
    9 − 5 = ___

4. 6 + 5 = ___
    5 + 6 = ___
    11 − 5 = ___
    11 − 6 = ___

Explain It • Daily Reasoning

How many facts are in the fact family for 12, 6, and 6?
Use 🎲 🎲 to prove your answer.

Chapter 14 • Practice Addition and Subtraction   two hundred thirty-one **231**

Practice and Problem Solving

Add or subtract.
Write the numbers in the fact family.

1.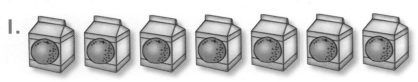

   7   5   12   12
   +5 +7 − 5 − 7

   12 ☐ ☐ ☐

2.

   6   4   10   10
   +4 +6 − 4 − 6

   ☐ ☐ ☐

3.

   9   3   12   12
   +3 +9 − 9 − 3

   ☐ ☐ ☐

Problem Solving
Logical Reasoning

Solve the riddle. Write the number.

4. When I am added to 4, we make 10.
   What number am I? _____

Write About It • Write a number sentence that shows your answer for Exercise 4. Then write the rest of the fact family.

HOME ACTIVITY • Tell your child an addition fact with a sum of 12 or less. Have your child tell the other number sentences in the fact family.

Name _____

Sums and Differences to 12

Learn

These are some ways to find sums and differences.

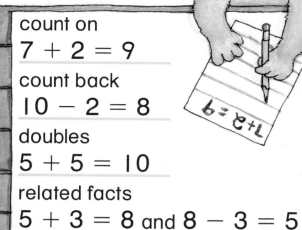

count on
7 + 2 = 9

count back
10 − 2 = 8

doubles
5 + 5 = 10

related facts
5 + 3 = 8 and 8 − 3 = 5

Check

Write the sum or difference.

1. 7 + 2 = __9__

2. 11 − 3 = ___
3. 6 − 0 = ___
4. 7 + 3 = ___

5. 10 − 2 = ___
6. 12 − 6 = ___
7. 5 + 3 = ___

8. 9 + 0 = ___
9. 9 − 2 = ___
10. 11 − 2 = ___

11. 8 − 4 = ___
12. 10 + 2 = ___
13. 6 + 3 = ___

14. 9 − 5 = ___
15. 8 − 8 = ___
16. 4 + 3 = ___

17. 11 − 5 = ___
18. 6 + 6 = ___
19. 4 + 5 = ___

Explain It • Daily Reasoning

How could knowing 12 − 3 = 9 help you find the difference for 12 − 4?

12 − 4

Practice and Problem Solving

Write the sum or difference.

1. 8
   +4
   ──
   12

2. 10
   − 3
   ──

3. 6
   +6
   ──

4. 12
   − 7
   ──

5. 3
   +8
   ──

6. 8
   −5
   ──

7. 9
   −4
   ──

8. 11
   − 2
   ──

9. 3
   +5
   ──

10. 12
    − 4
    ──

11. 5
    +5
    ──

12. 7
    −3
    ──

13. 6
    +2
    ──

14. 10
    − 6
    ──

15. 4
    +3
    ──

16. 11
    − 8
    ──

17. 3
    +7
    ──

18. 9
    +1
    ──

19. 5
    +7
    ──

20. 11
    − 6
    ──

21. 6
    +3
    ──

22. 10
    − 5
    ──

23. 7
    +4
    ──

24. 6
    +5
    ──

Problem Solving

Visual Thinking

25. Write the fact family that tells about the picture.

  Write About It • How would your fact family be different if there were 1 more cherry in the second row?

🏠 **HOME ACTIVITY** • With your child, make flash cards for the addition and subtraction facts through 12. Practice the facts each day.

234 two hundred thirty-four

Name _____

Algebra: Missing Numbers

Learn

What is the missing number?

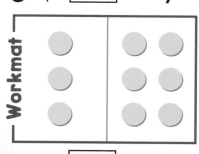

3 + ☐ = 9

3 + 6 = 9

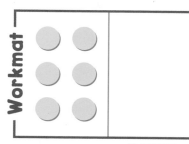

9 − 3 = ☐

9 − 3 = 6

Check

Write the missing number.
Use ● if you need to.

Use a related fact to help you.

1. 6 + ☐ = 10   10 − 6 = ☐

2. ☐ + 3 = 6   6 − 3 = ☐

3. 8 + ☐ = 12   12 − 8 = ☐

4. ☐ + 5 = 11   11 − 5 = ☐

5. 5 + ☐ = 12   12 − 5 = ☐

Explain It • Daily Reasoning

Use ● to show why you can solve ☐ + 5 = 9 by using subtraction.

Practice and Problem Solving

Write the missing number.
Use if you need to.

1. $2 + \boxed{5} = 7$ $7 - 2 = \boxed{5}$

2. $\square + 3 = 9$ $9 - 3 = \square$

3. $7 + \square = 10$ $10 - 7 = \square$

4. $\square + 7 = 8$ $8 - 7 = \square$

5. $8 + \square = 9$ $9 - 8 = \square$

6. $\square + 5 = 8$ $8 - 5 = \square$

7. $3 + \square = 11$ $11 - 3 = \square$

Problem Solving
Algebra

8. What is the missing number?

$$\begin{array}{r} 9 \\ +\blacksquare \\ \hline 11 \end{array} \qquad \begin{array}{r} \blacksquare \\ +9 \\ \hline 11 \end{array} \qquad \begin{array}{r} 11 \\ -9 \\ \hline \blacksquare \end{array} \qquad \begin{array}{r} 11 \\ -\blacksquare \\ \hline 9 \end{array}$$

■ = _____

Write About It • Look at Exercise 8.
Write the numbers that are in the fact family. Explain how you know.

HOME ACTIVITY • Put 12 small items in a bag. Have your child remove some, count them, and tell how many are left in the bag. Repeat.

Problem Solving
Choose a Strategy

You can use different ways to solve a problem.

Scott ate 4 hot dogs.
Tal also ate 4 hot dogs.
How many hot dogs did they eat in all?

Make a model.	Draw a picture.	Write a number sentence.
 		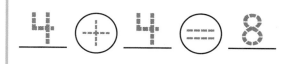
__8__ hot dogs	__8__ hot dogs	__8__ hot dogs

Choose a way to solve each problem.
Make a model 🧊, draw a picture 🖍,
or write a number sentence ✏️.
Show your work.

THINK: Which way do I want to solve this problem?

1. 8 ants are on the picnic table.
   3 go away.
   How many ants are there now?

   __5__ ants

2. The basket has 11 apples.
   Jose takes 2 apples.
   How many apples are there now?

   _____ apples

Chapter 14 • Practice Addition and Subtraction two hundred thirty-seven **237**

Problem Solving Practice

Choose a way to solve each problem.
Make a model 🔲, draw a picture 🖍, or write a number sentence ✏.
Show your work.

THINK: What is another way to solve the problem?

1. Meg's mom makes 12 muffins.
   Her family eats 9 muffins.
   How many muffins are left?

   _____ muffins

2. Sara sets out 9 plates.
   Children take 3 plates.
   How many plates are left?

   _____ plates

3. There are 4 pretzels.
   Hans brings 3 more pretzels.
   How many pretzels are there now?

   _____ pretzels

4. Rick eats 4 cherries.
   Peter eats 6 cherries.
   How many do they eat in all?

   _____ cherries

HOME ACTIVITY • Ask your child to explain how he or she solved each problem on this page. Then ask him or her to show a different way to solve each problem.

Name _____

Extra Practice

Write each sum or difference.
Circle the related facts.

1. $10 - 2 = $ _____ $9 + 3 = $ _____ $12 - 3 = $ _____

Add or subtract. Write the numbers in the fact family.

2. $7 + 4 = $ _____

   $4 + 7 = $ _____

   $11 - 4 = $ _____

   $11 - 7 = $ _____

3. $6 + 3 = $ _____

   $3 + 6 = $ _____

   $9 - 3 = $ _____

   $9 - 6 = $ _____

   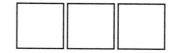

Write the sum or difference.

4. $\begin{array}{r} 8 \\ +1 \\ \hline \end{array}$ 5. $\begin{array}{r} 11 \\ -5 \\ \hline \end{array}$ 6. $\begin{array}{r} 12 \\ -4 \\ \hline \end{array}$ 7. $\begin{array}{r} 6 \\ +4 \\ \hline \end{array}$ 8. $\begin{array}{r} 3 \\ +4 \\ \hline \end{array}$ 9. $\begin{array}{r} 10 \\ -2 \\ \hline \end{array}$

Write the missing number.

10. ☐ $+ 5 = 11$ $11 - 5 = $ ☐

Problem Solving

Choose a way to solve the problem.

11. Lily has 8 cherries.
    Her friend gives her 3 cherries.
    How many does Lily have now?

    _____ cherries

Chapter 14 • Practice Addition and Subtraction

Name _____

✅ Review/Test

Concepts and Skills

Write each sum or difference.
Circle the related facts.

1. 4 + 7 = ___   11 − 7 = ___   11 − 2 = ___

Add or subtract. Write the numbers in the fact family.

2. 4 + 8 = ___
   8 + 4 = ___
   12 − 8 = ___
   12 − 4 = ___

3. 6 + 5 = ___
   5 + 6 = ___
   11 − 5 = ___
   11 − 6 = ___
   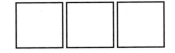

Write the sum or difference.

4. 9 +1   5. 12 − 6   6. 10 − 4   7. 7 +5   8. 4 +5   9. 11 − 3

Write the missing number.

10. 8 + ☐ = 12   12 − 8 = ☐

Problem Solving

Choose a way to solve the problem.

11. Alex has 12 pretzels.
    He gives 4 pretzels to friends.
    How many pretzels are left?

    _____ pretzels

240 two hundred forty

Name _____

★Standardized Test Prep
Chapters 1–14

Choose the answer for questions 1–5.

1. 12   18   17
   − 5   ○   ○
            8   7
            ○   ○

2.   5   1   2
   +6   ○   ○
           10   11
           ○   ○

3. Which is the missing number?

   6 + ☐ = 10   2   4   6   8
                    ○   ○   ○   ○

4. Gabe had 11 berries. He ate 4. Which number sentence tells how many berries Gabe has left?

   11 − 7 = 4   11 − 4 = 7   7 + 4 = 11   4 + 7 = 11
      ○   ○   ○   ○

5. Which is a way to make 9?

   4 + 2   5 + 4   4 + 1   4 + 3
     ○   ○   ○   ○

Show What You Know

6. Write four number sentences that are in the same fact family. Draw a picture of each number sentence to explain.

___○___○___   ___○___○___

___○___○___   ___○___○___

Chapter 14   two hundred forty-one **241**

IT'S IN THE BAG
Animals' Picnic Basket

PROJECT Create a slide-through picnic basket of addition and subtraction facts.

You Will Need
- Large brown bag
- Pattern tracer
- Cardboard facts strip
- Scissors
- Crayons

Directions

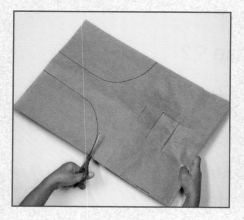

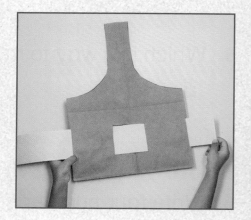

1. Put the bag flat in front of you. Place the pattern at the top of the bag. Trace around the pattern.

2. Cut on the lines you drew.

3. Put the end of the facts strip in the left side. Push it through to the other side.

4. Decorate your picnic basket. Draw and write addition and subtraction facts on the strip so they show in the basket's window.

The Animals' Picnic

BY DAVID McPHAIL

🏠 This book will help me review doubles.

This book belongs to _____.

All the animals were having a picnic.

One mouse invited one mouse.

They went by bike.

Two sheep invited two sheep.

They went by car.

Three rabbits invited three rabbits.

They went by wheelbarrow.

Four ants invited four ants.

They walked.

The ants got there first.

Name _____

PROBLEM SOLVING ON LOCATION

In Turkey Run State Park

You can canoe, hike, and camp outdoors at Turkey Run State Park in Indiana.

1 These tents hold 2 people. Skip count by twos. Write how many people the tents can hold.

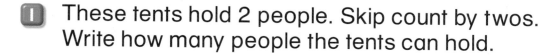

2 , 4 , ____ , ____ , ____ , ____ people

2 These tents hold 5 people. Skip count by fives. Write how many people the tents can hold.

5 , ____ , ____ , ____ , ____ , ____ , ____ , ____ people

3 These tents hold 10 people. Skip count by tens. Write how many people the tents can hold.

10 , ____ , ____ , ____ , ____ , ____ , ____ people

Unit 3 • Chapters 9–14

CHALLENGE

Represent Numbers in Different Ways

This number is shown in three different ways.

11 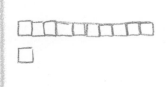 10 ⊕ 1

Show each number three different ways.

Number	Use ▪. Draw to show.	Draw a picture.	Write a name for the number.
1. 7			__ ◯ __
2. 10			__ ◯ __
3. 8			__ ◯ __
4. 12			__ ◯ __

Name _____

✓ Study Guide and Review

Vocabulary

Write how many **tens** and **ones**.
Write the number in a different way.

1.

   _____ tens _____ ones = _____

   _____ + _____

2.

   _____ tens _____ ones = _____

   _____ + _____

Skills and Concepts

Circle the number that is greater.
Write the numbers.

3.

   _____ is greater than _____.

   _____ > _____

4.

   _____ is greater than _____.

   _____ > _____

Circle the number that is less.
Write the numbers.

5.

   _____ is less than _____.

   _____ < _____

6.

   _____ is less than _____.

   _____ < _____

Unit 3 • Study Guide and Review

Sort. Draw to complete the picture graph.

7.

Lunches We Like						
hot dog						
pizza						
soup						

Count by fives. Write the missing numbers.

8. 5, 10, ____, ____, 25, ____, ____, 40

Add or subtract. Think of a related fact to help.

9. 8   10. 10   11. 9   12. 5   13. 6   14. 11
   +4   −5   +3   −5   +6   −2

Problem Solving

Write the number sentence.
Draw a picture to check.

15. There are 12 eggs.
    Dad cooks 6 of them.
    How many eggs are left?

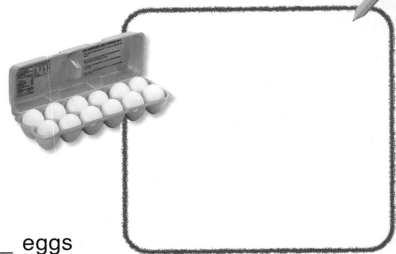

____ ◯ ____ ◯ ____ eggs

246 two hundred forty-six

✓ Performance Assessment

Buying Fish

Megan and her sister Erin bought some fish at the pet store.

- Erin bought 1 more fish than Megan.

- Megan bought fewer than 7 fish.

Write a doubles-plus-one sentence that fits this math story.

Show your work.

Name ________________________________

TECHNOLOGY

Calculator • Skip Counting

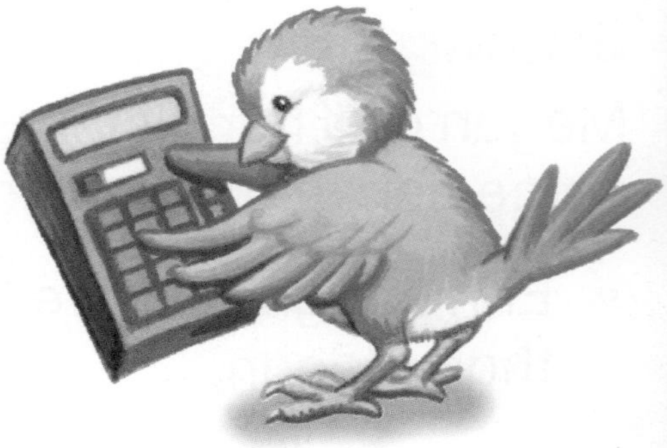

You can use a calculator to skip count.

Skip count by twos.

Press ON/C + 2 =.

Read the number 2.

Press =.

Read the number 4.

Press =.

Read the number 6.

So, count 2, 4, 6 to count by twos.

Practice and Problem Solving

Use a calculator.

1. Skip count by threes.

Press ON/C + 3 = ☐ = ☐ = ☐ = ☐ = ☐ = ☐

2. Count by adding the next larger number each time.

Press ON/C 1 + 2 = ☐ + 3 = ☐ + 4 = ☐

TECHNOLOGY

LOOKING BACK SCHOOL HOME CONNECTION

Dear Family,

In Unit 3 we learned about graphs, numbers to 100, and facts to 12. Here is a game for us to play together. This game will give me a chance to share what I have learned.

Love,

_____

Directions
1. Cover each apple with a penny.
2. Pick up 1 penny.
3. Use the number on that apple. Tell an addition or subtraction fact that uses doubles or doubles plus one.
4. Put the penny on the graph to show the kind of fact you made.
5. Take turns until a row is full. Look at the graph. Count the number of doubles facts and doubles plus one facts.
6. Tell which kind of fact you made more of. Play again.

Materials
21 pennies or beans

Dear Family,

During the next few weeks, we will learn about solid figures, plane shapes, and patterns. We will also learn more about addition and subtraction to 20. Here is important math vocabulary and a list of books to share.

Love,

_____

Vocabulary

rectangle
square
circle
triangle
sphere
cone
cylinder
pyramid
cube
rectangular prism

Vocabulary Power

Solid Figures:

rectangular prism   sphere

cone   cylinder

pyramid   cube

Plane Shapes:

rectangle   square   circle   triangle

BOOKS TO SHARE

To read about geometry and patterns with your child, look for these books in your library.

When a Line Bends... A Shape Begins, by Rhonda Gowler Greene, Houghton Mifflin, 2001.

Circus Shapes, by Stuart J. Murphy, HarperCollins, 1998.

The Very Busy Spider, by Eric Carle, Penguin Putnam, 1999.

A Fair Bear Share, by Stuart J. Murphy, HarperCollins, 1998.

 Visit *The Learning Site* for additional ideas and activities. www.harcourtschool.com

CHAPTER 15
Solid Figures and Plane Shapes

FUN FACTS

Most sand castles are built from a pile of sand that has been packed down. Then molds can be placed on top.

Theme: Shapes in Our World

Name _____

✓ Check What You Know

Sort Solid Figures

Color the shape blue.

Color the shape red.

Color the shape yellow.

Color the shape green.

1.

2.

3.

4.

5.

6.

Sort Plane Shapes

Circle the same shape.

7.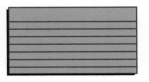

8.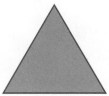

250 two hundred fifty   Use this page to review important skills needed for this chapter.

Name _____

Solid Figures

Explore (Hands On)

Vocabulary
- cylinder
- pyramid
- rectangular prism
- sphere
- cone
- cube

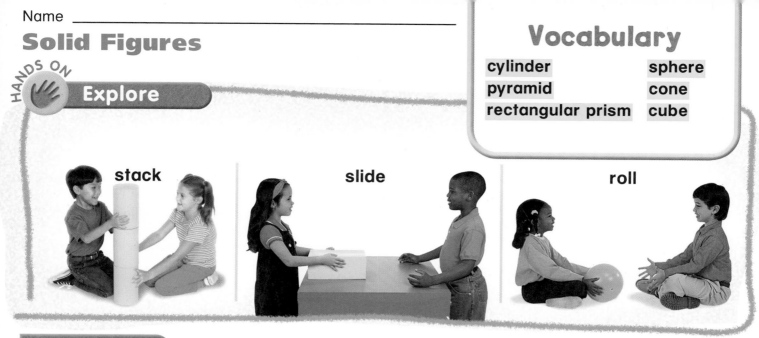

stack | slide | roll

Connect

Use solids. Sort. Write **yes** or **no**.

		Does it stack?	roll?	slide?
1.	sphere	no		
2.	cone			
3.	cube			
4.	cylinder			
5.	pyramid			
6.	rectangular prism			

Explain It • Daily Reasoning

Which solids roll? Tell why.

Chapter 15 • Solid Figures and Plane Shapes

Practice and Problem Solving

cube   pyramid   rectangular prism   sphere

These can slide.   This is round and curved. It rolls.

Use solids.

1. Color each solid that will stack.

2. Color each solid that will roll.

3. Color each solid that will slide.

Problem Solving
Logical Reasoning

4. Cross out the solid that does not belong. Circle the sentence that tells why.

It will not stack.

It will not roll.

 Write About It • Draw something in your classroom that will roll. Tell why.

HOME ACTIVITY • Find objects that are shaped like the solids on this page. Work with your child to find out which ones will stack, roll, and slide.

Open and Closed

Vocabulary
open figure
closed figure

Learn

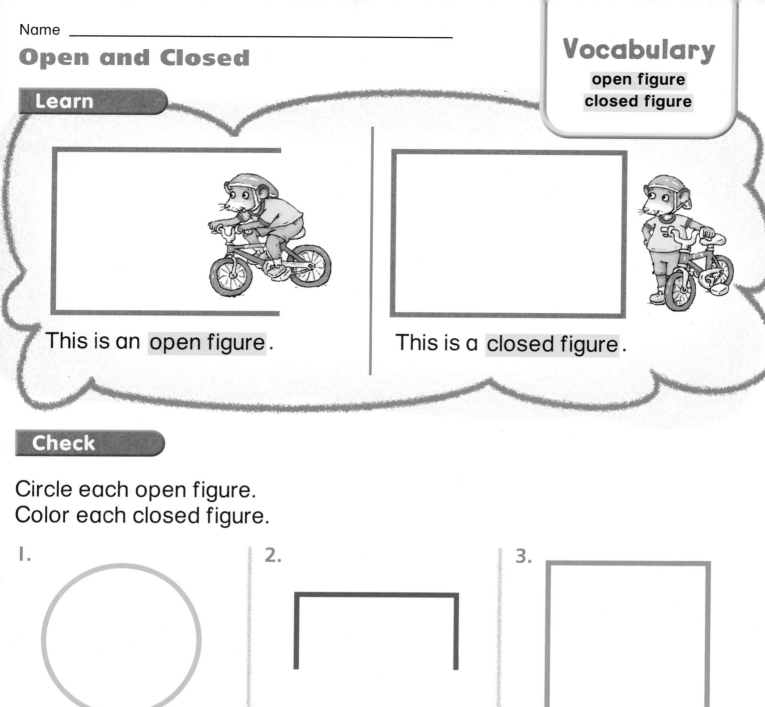

This is an open figure.

This is a closed figure.

Check

Circle each open figure.
Color each closed figure.

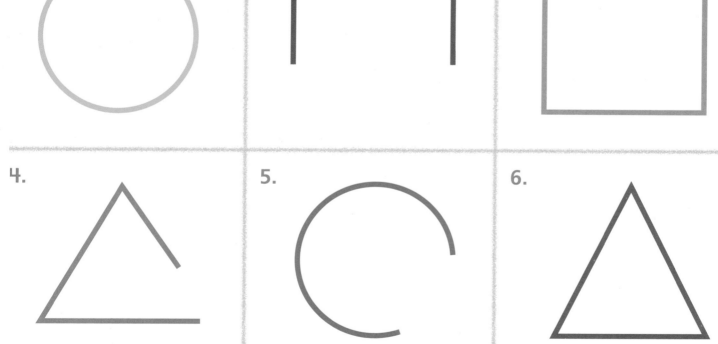

Explain It • Daily Reasoning

How are open and closed figures different?

Practice and Problem Solving

Circle each open figure.
Color each closed figure.

1.
2.
3.

4.
5.
6.

7.
8.
9.

Problem Solving
Application

10. Draw a face. Use only closed figures.

Write About It • Look at Exercise 10.
Which closed figures did you use?
Write about the face you drew.

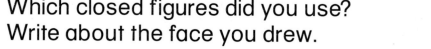
HOME ACTIVITY • Have your child find three objects at home whose outlines are closed figures, for example, the outline of a window.

268 two hundred sixty-eight

Name _____

Problem Solving Skill
Use a Picture

Vocabulary
above   below
close by   over
near   far
next to   beside
to the left of
to the right of

Follow the directions.

1. The [School] is **above** the .

    Draw a ☀ **above** the .

2. The ⚽ is **below** the .

    Draw a 🌼 **below** the .

3. The is **close by** the .

    Draw a 🚗 **close by** the .

4. The is **over** the .

    Draw a 🐦 **over** the .

Chapter 16 • Spatial Sense

Problem Solving Practice

Follow the directions.

1. The is near the .

   Draw a near the 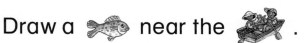.

2. The is far from the .

   Draw a far from the .

3. The is next to the .

   Draw a next to the 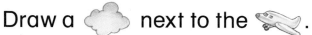.

4. The is beside the .

   Draw a beside the 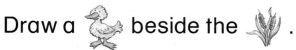.

5. The is to the left of

   the .

   Draw a to the left of

   the .

6. The is to the right of

   the .

   Draw a to the right of

   the .

HOME ACTIVITY • Have your child use the words *far*, *next to*, and *beside* to describe the positions of objects in a room in your home. Then have your child use *above*, *below*, *over*, and *near*.

Name _____

Give and Follow Directions

Vocabulary
up   left
down   right

Learn

From **Start**, go right 3.
Go up 2.
Where are you?

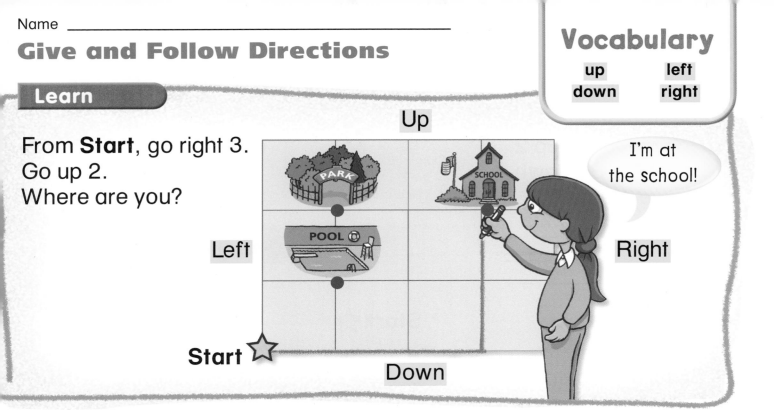

Check

Follow the directions in order.
Draw the path. Write the place.

1. Go down 1. Go left 4.
   Where are you?

   _____

2. Go right 2. Go up 1. Go right 2.
   Go up 1. Where are you?

   _____

Explain It • Daily Reasoning

How could you go from the
playground to the zoo? Is there
more than one way? Explain.

Chapter 16 • Spatial Sense

Practice and Problem Solving

Follow the directions in order.
Draw the path. Write the place.

1. Go right 3. Go up 2.
   Go left 1. Go down 1.
   Where are you?

   _____

2. Go left 4. Go down 1.
   Go right 3. Go down 1.
   Where are you?

   _____

Problem Solving
Visual Thinking

3. Help the puppy get to the bones.
   Write **up**, **down**, **left**, or **right**.

   Go _____ 3.

   Go _____ 2.

   Go _____ 2.

 Write About It • Choose a place in your school. Make a map. Tell how to get there. Then draw the path.

🏠 **HOME ACTIVITY** • Ask your child to tell another way the puppy could get to the bones in Exercise 3.

Name _____

Symmetry

Vocabulary
line of symmetry

Hands On Explore

1. Fold your paper.

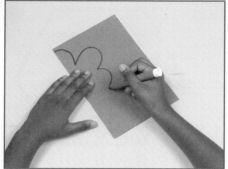
2. Start at the fold. Draw a shape.

3. Cut along the line.

4. Open your shape.

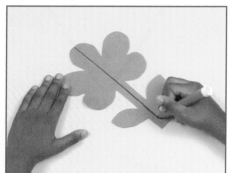
5. Draw a line down the middle.

The line down the middle is called a **line of symmetry**. It shows two parts that match.

Connect

Draw a line of symmetry to show two matching parts.

1.

2.

Explain It • Daily Reasoning

Fold a shape on the line of symmetry. How can you tell if the two parts match?

Chapter 16 • Spatial Sense

Practice and Problem Solving

Draw a line of symmetry to show two matching parts.

1. 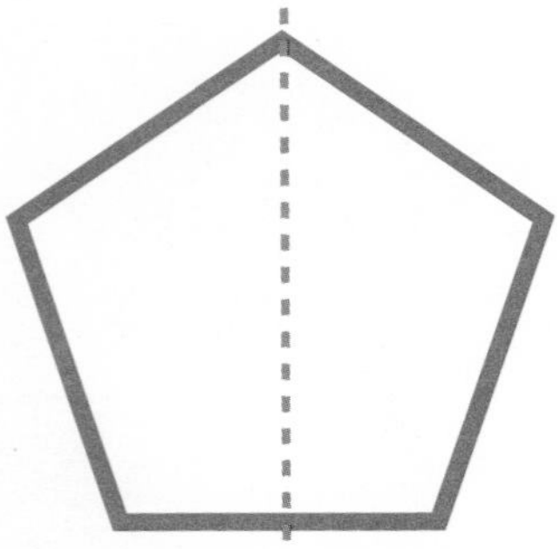
2. 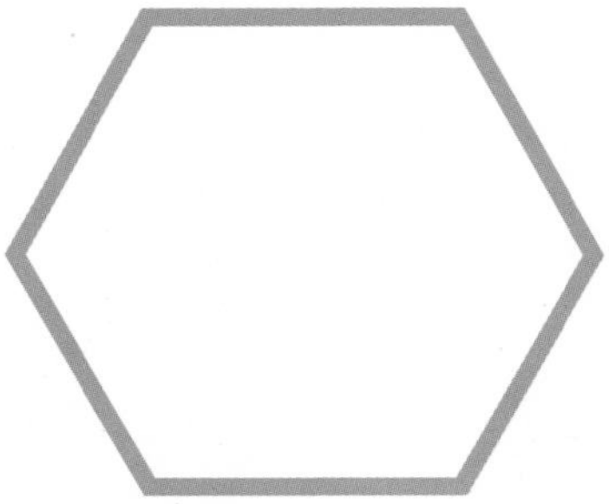
3. 
4. 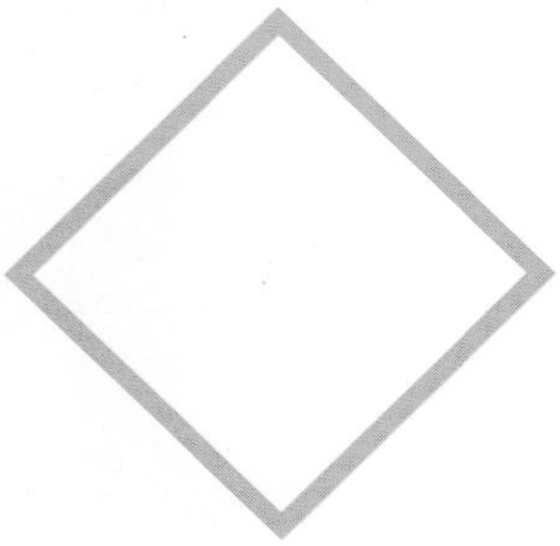
5. 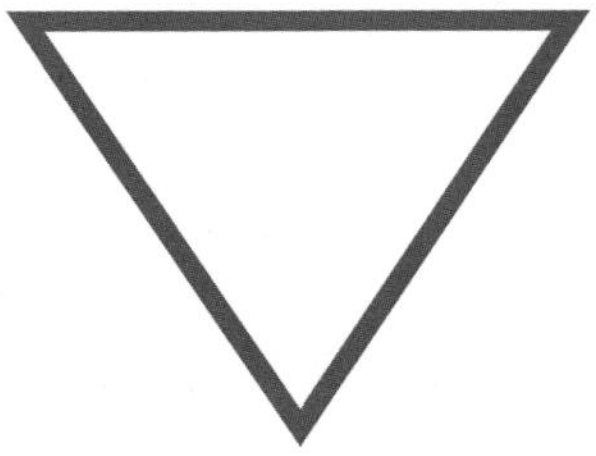
6.
7. 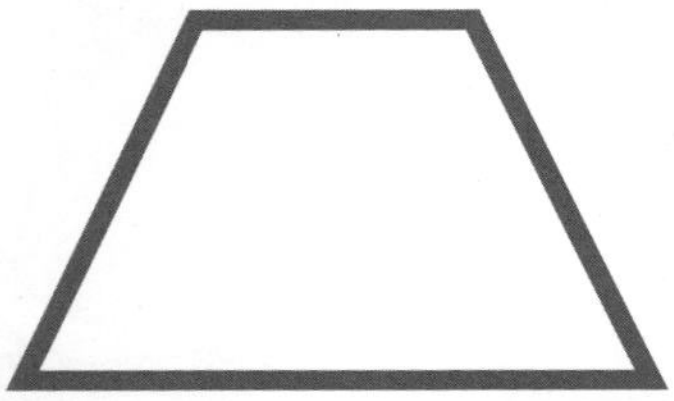
8.
9.

Problem Solving
Visual Thinking

10. Draw a different line of symmetry on each square.

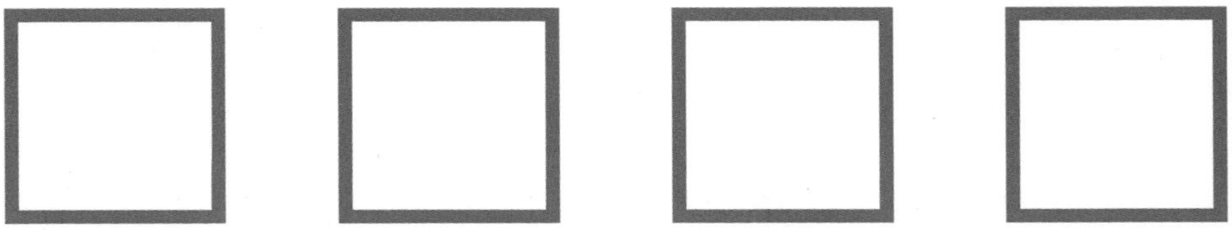

Write About It • Look at Exercise 10. Hold a mirror on each line of symmetry. Write about what you see.

HOME ACTIVITY • Find objects that are symmetrical. Have your child trace the line of symmetry of each with a finger.

Algebra: Make New Patterns

Explore

Use the same shapes to make a different pattern. Draw your new pattern.

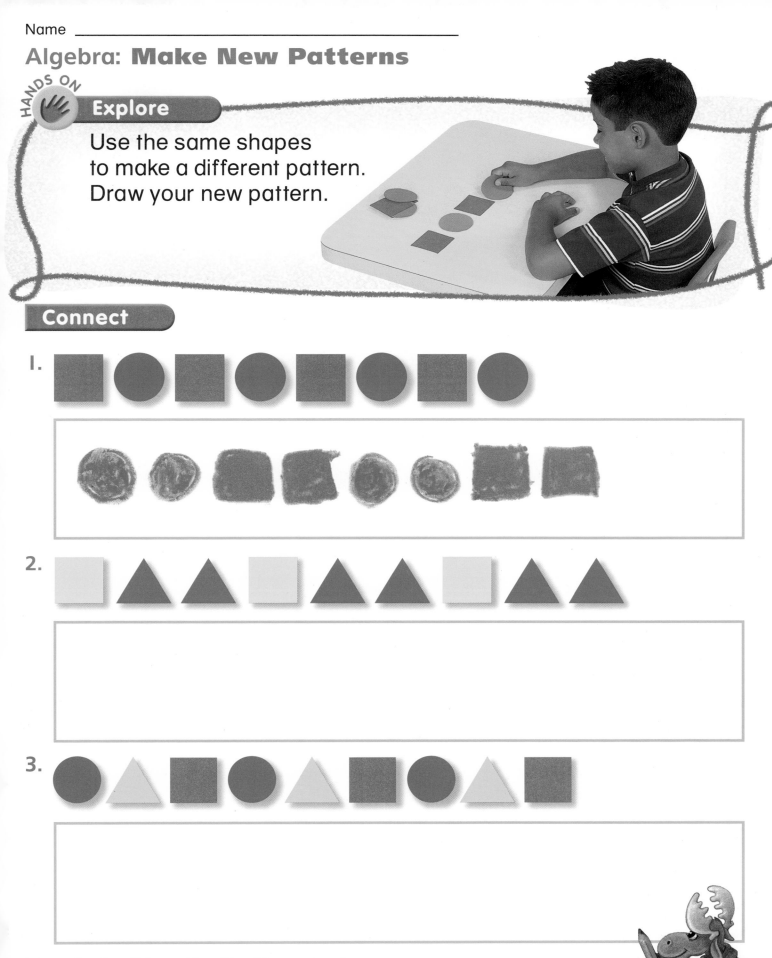

Connect

1.

2.

3.

Explain It • Daily Reasoning

How are your patterns the same as the ones shown? How are they different?

Chapter 17 • Patterns

Practice and Problem Solving

Use the same shapes to make a different pattern. Draw your new pattern.

1.

2.

3.

Problem Solving
Logical Reasoning

4. Find the pattern. Draw what comes next.

 Write About It • Use ➡ → .
Make your own pattern.

HOME ACTIVITY • Have your child arrange objects in a pattern and then explain the pattern to you.

Problem Solving Skill
Correct a Pattern

Find the mistake in the pattern.

Each pattern unit is 3 shapes long.
Find the pattern.
Circle the mistake. Draw the correct shape.

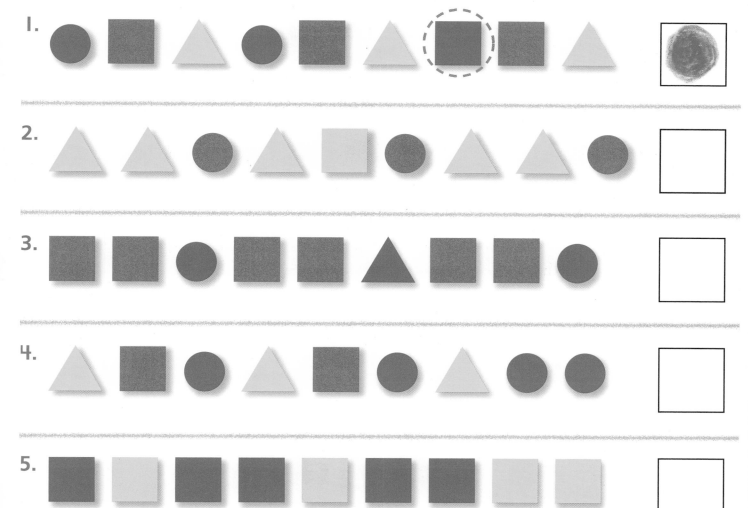

Problem Solving Practice

Find the pattern.
Circle the mistake.
Draw the correct shape.

Each pattern unit is 3 shapes long.

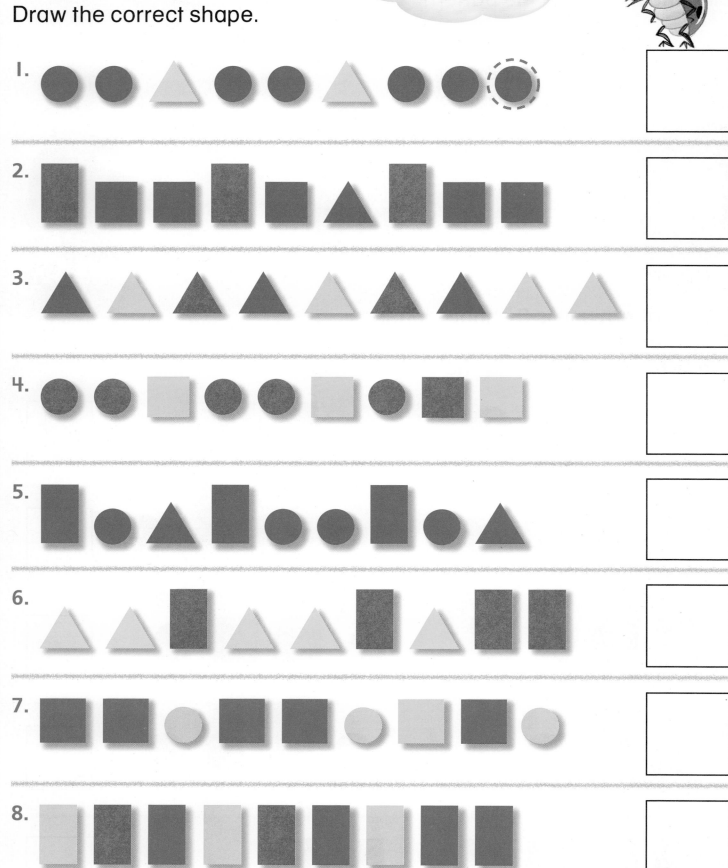

Problem Solving Skill
Transfer Patterns

You can show the same pattern in a different way.

Use shapes to show the same pattern.
Draw the shapes.

1.

2.

3.

Chapter 17 • Patterns

Problem Solving Practice

Use shapes to show the same pattern.
Draw the shapes.

1.

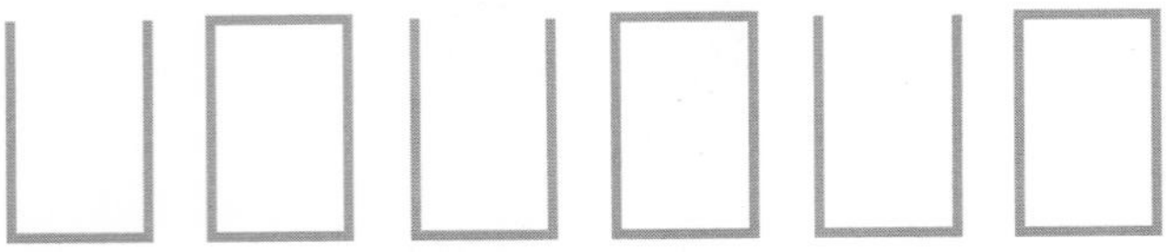

2.

3.

4.

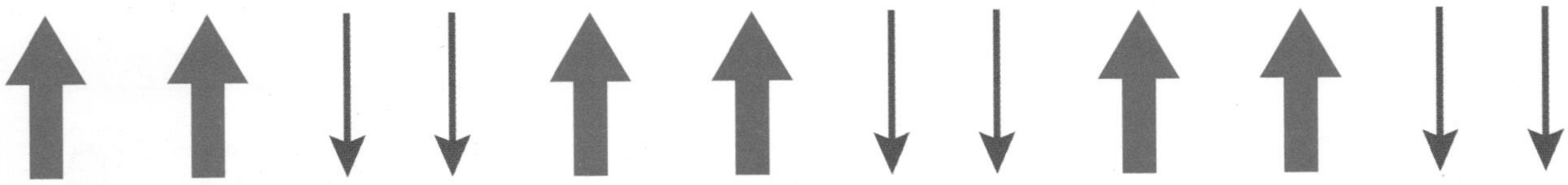

HOME ACTIVITY • Arrange objects in a pattern. Have your child use different objects to show the same pattern.

Name _____

Extra Practice

Find the pattern. Then color to continue it.

1.    2.    3.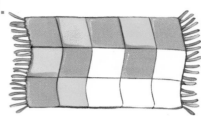

4. Circle the pattern unit.

5. Use the same shapes to make a different pattern. Draw your new pattern.

Problem Solving

Find the pattern.
Circle the mistake. Draw the correct shape.

Each pattern unit is 3 shapes long.

6.

7.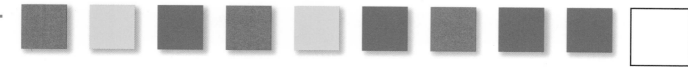

Chapter 17 · Patterns

two hundred ninety-three **293**

Name _____

✅ Review/Test

Concepts and Skills

Find the pattern. Then color to continue it.

1.
2.
3.

4. Circle the pattern unit.

5. Use the same shapes to make a different pattern. Draw your new pattern.

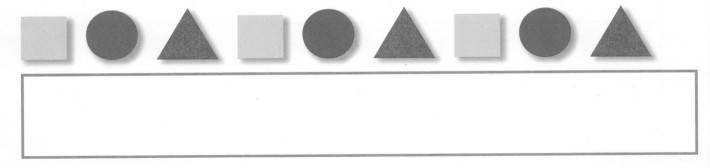

Problem Solving

6. Use shapes to show the same pattern. Draw the shapes.

294 two hundred ninety-four

Name _____

★Standardized Test Prep
Chapters 1–17

Choose the answer for questions 1 – 3.

1. Which shows the right way to continue the pattern?

2. Find the pattern unit. Then find the mistake. Which shape will correct the mistake?

3. How many faces are on a ▭ ?

   2   5   6   7
   ○   ○   ○   ○

Show What You Know

4. Use different shapes to show the same pattern. Draw and explain your new pattern.

MATH GAME

Pattern Play

Play with a partner.

1. Spin the 🔄. Take any Attribute Links that match that shape or color.

2. Take turns until each player has three links.

3. Use these to make a pattern unit of shapes or colors. Draw your pattern unit. You will make your pattern on the table.

4. Spin again. If that shape or color is in your unit, use it in your pattern where you can.

5. The first player to repeat his or her pattern unit two times wins.

You will need

Attribute Links

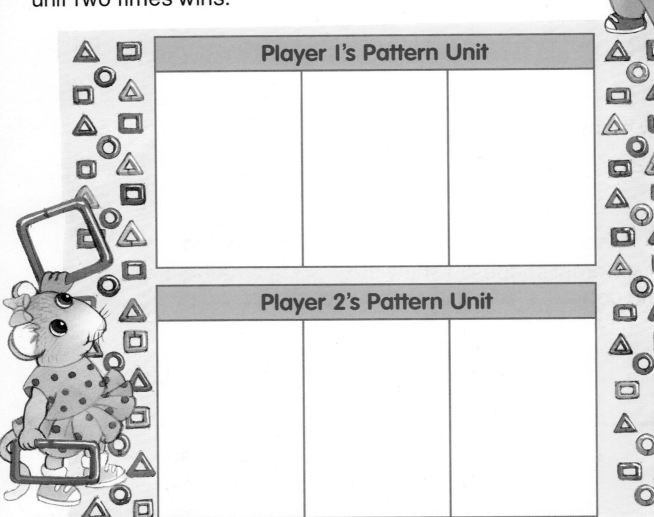

Player 1's Pattern Unit

Player 2's Pattern Unit

CHAPTER 18
Addition Facts and Strategies

FUN FACTS

Each paw has 5 toes and 5 claws.

Theme: Wildlife

Name _____

✓ Check What You Know

Count On to Add

Circle the greater number.
Use the number line. Count on to add.

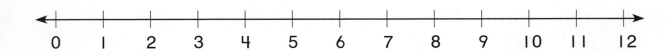

1. 8 +2
2. 3 +6
3. 1 +6
4. 7 +3
5. 2 +9
6. 2 +7

Doubles and Doubles Plus 1

Write the three sums.
Then circle the doubles fact.

7. 2 +2   2 +3   3 +2
8. 4 +4   4 +5   5 +4

Add 3 Numbers

Circle the two numbers you add first.
Write the sum.

9. 2 4 +4
10. 3 1 +6
11. 4 3 +4
12. 3 0 +5
13. 6 2 +1
14. 5 5 +2

298 two hundred ninety-eight   Use this page to review important skills needed for this chapter.

Name _____

Doubles and Doubles Plus 1

Vocabulary
doubles
doubles plus one

 Explore

```
  8
 +8
 ---
 16
```

8 + 8 = 16 is a **doubles** fact.
8 + 9 = 17 is a **doubles plus one** fact.

```
  8
 +9
 ---
 17
```

Connect

Use ◯. Write the sums.

1. 5   5   2. 2   3   3. 7   7
   +5   +6   +2   +2   +7   +8

4. 0   0   5. 6   7   6. 3   3
   +0   +1   +6   +6   +3   +4

7. 4   4   8. 8   9   9. 9   9
   +4   +5   +8   +8   +9   +10

Explain It • Daily Reasoning

What are two ways you could find the sum for 10 + 10?

Chapter 18 • Addition Facts and Strategies   two hundred ninety-nine **299**

Practice and Problem Solving

Write the sums.

1. $4 + 4 = \underline{8}$, so $5 + 4 = \underline{9}$

2. $7 + 7 = \underline{}$, so $7 + 8 = \underline{}$

3. $5 + 5 = \underline{}$, so $6 + 5 = \underline{}$

4. $9 + 9 = \underline{}$, so $9 + 10 = \underline{}$

5. $1 + 1 = \underline{}$, so $1 + 2 = \underline{}$

6. $3 + 3 = \underline{}$, so $4 + 3 = \underline{}$

7. $8 + 8 = \underline{}$, so $8 + 9 = \underline{}$

Problem Solving

Algebra

Write the missing numbers.

8. $\begin{array}{r}5\\+\square\\\hline 10\end{array}$ $\begin{array}{r}5\\+\square\\\hline 11\end{array}$

9. $\begin{array}{r}2\\+\square\\\hline 4\end{array}$ $\begin{array}{r}2\\+\square\\\hline 5\end{array}$

10. $\begin{array}{r}6\\+\square\\\hline 12\end{array}$ $\begin{array}{r}6\\+\square\\\hline 13\end{array}$

 Write About It • Look at Exercises 8, 9, and 10. Explain the pattern you see.

HOME ACTIVITY • Have your child tell you the doubles facts and the doubles plus one facts for 6, 7, 8, and 9 (6 + 6 = 12, 6 + 7 = 13; 7 + 7 = 14, 7 + 8 = 15; and so on).

Name _____

10 and More

Explore

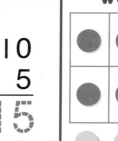

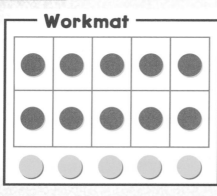

10
+ 5

15

I can use a ten frame to show 10 + 5.

Connect

Use ◯ and Workmat 7 to add.
Draw the ◯. Write the sum.

1. 10
   + 7

    17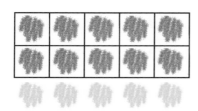

2. 10
   + 3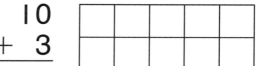

3. 10
   + 8

4. 10
   + 4

5. 10
   + 6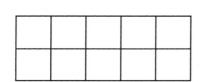

6. 10
   + 2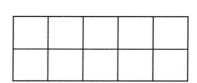

Explain It • Daily Reasoning

What happens when you add 10 to any number less than 10?

Chapter 18 • Addition Facts and Strategies

Practice and Problem Solving

Write the sum.

1. 10
   + 9

   19

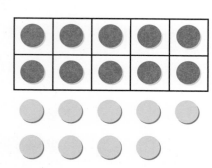

2. 10
   + 4

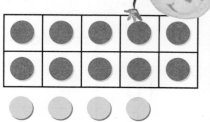

3. 10
   + 1

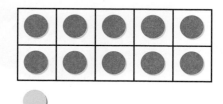

4. 10
   + 6

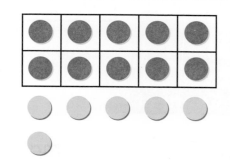

5. 10   6. 10   7. 10   8. 10   9. 10   10. 10
   + 2   + 7   + 3   + 5   + 8   + 0
   ---   ---   ---   ---   ---   ---

Problem Solving
Logical Reasoning

Choose a way to solve.

11. Jan plants 18 flowers in two rows. She plants 10 in the first row. How many are in the second row?

_____ flowers

Write About It • Look at Exercise 11. Explain how a row of 10 helped you.

HOME ACTIVITY • Ask your child to tell the sums for 10 + 1 through 10 + 9 (10 + 1 = 11, 10 + 2 = 12, and so on).

Name _____

Make 10 to Add

Vocabulary
make a ten

Explore Find the sum for 9 + 5.

Show 9.
Then show 5.

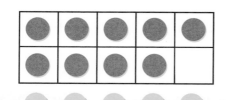

9
+5
―――
14

Make a ten.
Move 1 counter into the ten frame.

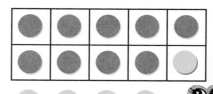

10
+4
―――
14

Connect

Use ⬤ and Workmat 7.
Show the numbers and add. Then make a ten and add.

1. 9
   +7

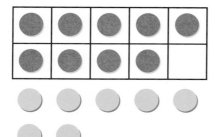

   10
   + 6

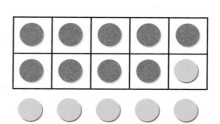

2. 9
   +4

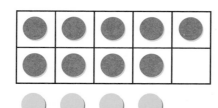

   10
   + 3

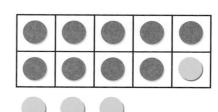

3. 9
   +6

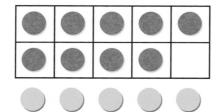

   10
   + 5

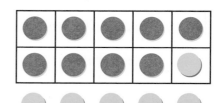

Explain It • Daily Reasoning

How do you know that 9 + 3 = 10 + 2?
Use ⬤ to prove your answer.

9 + 3   10 + 2

Chapter 18 • Addition Facts and Strategies   three hundred three **303**

Practice and Problem Solving

Use ◯ and Workmat 7.
Show the numbers and add. Then make a ten and add.

1. 9
   +9
   ——
   18
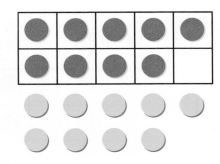

   10
   + 8
   ——
   18
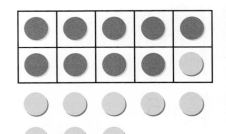

2. 9
   +2
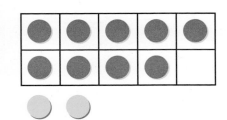

   10
   + 1
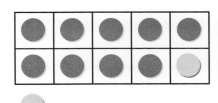

3. 9
   +8
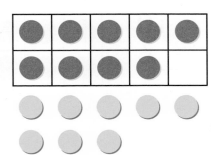

   10
   + 7
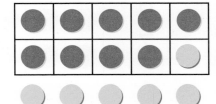

Problem Solving
Mental Math

Solve in your head.
Draw a picture to check.

4. There were 9 pine cones on the ground. Then 5 more pine cones fell. How many pine cones were on the ground then?

   _____ pine cones

 Write About It • Look at Exercise 4.
Explain how to make a ten to add 9 + 5.

⬠ **HOME ACTIVITY** • Ask your child to tell the sums for 9 + 1 through 9 + 9 (9 + 1 = 10, 9 + 2 = 11, and so on).

Name _____

Use Make a 10

Explore

Find the sum for 7 + 4.

First show 7. Then show 4.
Fill up the ten frame to add.

```
  7
+ 4
───
 11
```

You made a ten and have 1 extra.
10 + 1 = 11

Connect

Use ◯ and Workmat 7 to add.
Start with the greater number.
Draw the ◯. Write the sum.

1.  ```
     5
   + 8
   ───
    13
   ```
   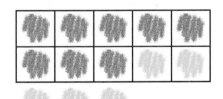

2.  ```
     7
   + 6
   ───
   ```

3.  ```
     8
   + 4
   ───
   ```
   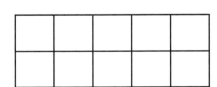

4.  ```
     5
   + 6
   ───
   ```
   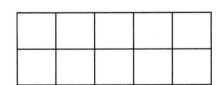

5.  ```
     8
   + 7
   ───
   ```

6.  ```
     7
   + 5
   ───
   ```
   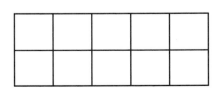

Explain It • Daily Reasoning

How do you make a ten to add two numbers?

Chapter 18 • Addition Facts and Strategies   three hundred five **305**

Practice and Problem Solving

Use ⬤ and Workmat 7 to add.
Start with the greater number.

THINK:
6 + 8 = 10 + 4

1.   6
   +8
   ──
   14

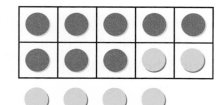

2. 9   3. 3   4. 5   5. 8   6. 8   7. 9
   +6   +8   +7   +9   +5   +7
   ──   ──   ──   ──   ──   ──

8. 4   9. 8   10. 3   11. 4   12. 5   13. 9
   +7   +6   +9   +8   +9   +4
   ──   ──   ──   ──   ──   ──

14. 6   15. 7   16. 7   17. 2   18. 6   19. 6
   +7   +8   +9   +9   +5   +8
   ──   ──   ──   ──   ──   ──

Problem Solving
Logical Reasoning

Choose a way to solve.

20. Josh needs to plant 15 seeds. He has 8 seeds. How many more seeds does he need?

_____ more seeds

Write About It • Look at Exercise 20.
Complete the number sentence. 8 + ☐ = 10 + 5

HOME ACTIVITY • Ask your child to read a problem on this page and tell how to solve it by making a ten. For example, solve 8 + 4 by making it 10 + 2.

Name _____

Algebra: Add 3 Numbers

Learn

You can add three numbers in any order.

⑦
③
+4

14

7
③
+④

14

You can make a ten.
7 + 3 = 10
10 + 4 = 14

You can use doubles.
3 + 4 = 7
7 + 7 = 14

Check

Circle the numbers you add first.
Write the sum.

1. 2
    7
   +3
   ___

2. 8
    4
   +4
   ___

3. 9
    3
   +1
   ___

4. 8
    3
   +3
   ___

5. 4
    7
   +3
   ___

6. 6
    5
   +4
   ___

7. 2
    2
   +7
   ___

8. 2
    5
   +5
   ___

9. 4
    2
   +6
   ___

10. 8
    8
   +1
   ___

11. 8
    2
   +7
   ___

12. 4
    3
   +4
   ___

Explain It • Daily Reasoning

Look at Exercise 1. Add two different numbers first.
Did the sum change? Why or why not?

Practice and Problem Solving

Circle the numbers you add first. Write the sum.

1. 1
   (5)
   +(5)

   11

2. 8
    2
   +1
   ---

3. 6
    3
   +7
   ---

4. 1
    2
   +9
   ---

5. 8
    2
   +6
   ---

6. 1
    7
   +7
   ---

7. 4
    9
   +6
   ---

8. 9
    6
   +1
   ---

9. 3
    6
   +3
   ---

10. 4
    4
   +2
   ---

11. 8
    4
   +2
   ---

12. 3
    5
   +5
   ---

13. 6
    1
   +6
   ---

14. 8
    1
   +9
   ---

15. 5
    7
   +3

Problem Solving
Application

Draw a picture to solve.

16. Kim picks 3 pink flowers. Tod picks 7 yellow flowers. Chris picks 5 purple flowers. How many flowers in all do the children pick?

    _____ flowers

 Write About It • Look at Exercise 16. Explain how you could make a ten. How could that help you solve the problem?

🏠 HOME ACTIVITY • Have your child use pennies to show how to add three numbers.

Name _____

Problem Solving Skill
Use Data from a Table

This table tells how many animals children saw at camp.

Animals	Number
chipmunks	6
rabbits	2
squirrels	4
deer	3

Use the table to answer the questions. Write a number sentence to solve.

1. How many deer and squirrels did they see?

   __7__ deer and squirrels   __3__ ⊕ __4__ ⊖ __7__

2. How many more chipmunks than deer did they see?

   _____ more chipmunks   ___ ◯ ___ ◯ ___

3. How many more deer than rabbits did they see?

   _____ more deer   ___ ◯ ___ ◯ ___

4. How many chipmunks and rabbits did they see?

   _____ chipmunks and rabbits   ___ ◯ ___ ◯ ___

5. How many small animals did they see in all?

   Find the numbers for small animals.

   _____ small animals   ___ ◯ ___ ◯ ___ ◯ ___

Chapter 18 • Addition Facts and Strategies   three hundred nine **309**

Problem Solving Practice

This table tells how many birds children saw at camp.

Birds	Number
owl	1
robins	5
blue jays	4
blackbirds	7

Use the table to answer the questions. Write a number sentence to solve.

1. How many robins and blue jays did they see in all?

    _____ robins and blue jays   ___○___○___

2. How many more robins than owls did they see?

    _____ more robins   ___○___○___

3. How many blackbirds and owls did they see in all?

    _____ blackbirds and owls   ___○___○___

4. How many more blackbirds than robins did they see?

    _____ more blackbirds   ___○___○___

5. How many birds did they see that were not blue?

    Find the numbers for birds that are not blue.

    _____ birds   ___○___○___○___

HOME ACTIVITY • Ask your child to explain how he or she solved each problem.

Name _____

Extra Practice

Write the sum.

1. 3   3
   +3   +4

2. 6   6
   +6   +7

3. 5   5
   +5   +6

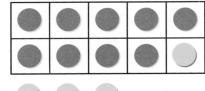

4.   9
            +4

5.   9
            +6

6. 8   7. 8   8. 2   9. 7   10. 6   11. 4
   +4   +8   +7   +7   +4   +5

12. 4   13. 7   14. 2   15. 4   16. 8   17. 3
    +4   +8   7   2   6   7
                      +3   +6   +1   +3

Problem Solving

This table tells how many animals children saw.
Use the table to answer the question.

18. How many animals did the children see in all?

_____ animals

Animals	Number
rabbits	3
deer	7
squirrels	4

Chapter 18 • Addition Facts and Strategies

Name _____

✓ Review/Test

Concepts and Skills
Write the sum.

1. 4   4
   +4   +5

2. 8   8
   +8   +9

3. 7   8
   +7   +7

4. 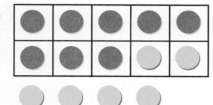 8
                         +6

5. 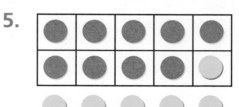 9
                         +7

6. 9   7. 6   8. 5   9. 9   10. 7   11. 5
   +5   +6   +5   +9   +5   +6

12. 6   13. 6   14. 7   15. 3   16. 9   17. 4
    +7   +9   2   4   7   8
                    +2   +7   +1   +4

Problem Solving

This table tells how many animals children saw.
Use the table to answer the question.

18. How many animals did the children see in all?

    ___ ○ ___ ○ ___ ○ ___

    _____ animals

Animals	Number
chipmunks	6
rabbits	6
squirrels	4

Name _____

★Standardized Test Prep
Chapters 1-18

Choose the answer for questions 1-5.

1. $7 + 7 = 14$, so $7 + 8 = \underline{}$

   14 ○   15 ○   16 ○   17 ○

2. Which shows the sum of $10 + 9$?

   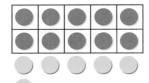   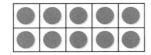   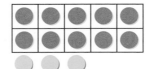   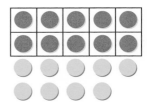

   ○   ○   ○   ○

3. The table shows how many vegetables the children picked.

   How many more cucumbers than beans did they pick?

   2 ○   4 ○   5 ○   8 ○

Vegetables Picked	
beans	8
cucumbers	12
tomatoes	5

4.
   $$\begin{array}{r} 9 \\ 1 \\ +3 \\ \hline \end{array}$$

   10 ○   12 ○   13 ○   15 ○

5. Which is the missing number?

   $8 + \boxed{} = 13$

   3 ○   5 ○   7 ○   8 ○

Show What You Know

6. Use the ten frame. Draw counters to explain how to make a ten to find the sum.

   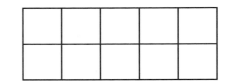

MATH GAME

Ten Plus

Play with a partner.

1. Put your ♟ at START.
2. Stack the [2] face down.
3. Start with 9. Take one number card.
4. Find the sum by making 10 first.
5. Say the number you add to 10 to get the sum.
6. Move your ♟ that many spaces.
7. The first player to get to END wins.

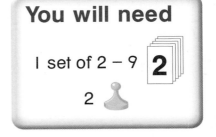

You will need

1 set of 2 – 9 [2]

2 ♟

CHAPTER 19 Subtraction Facts and Strategies

FUN FACTS

Hibiscus grows in many shades of its 7 basic colors: red, yellow, blue, pink, white, purple, and orange.

Theme: In the Garden

Name _____

✓ Check What You Know

Count Back to Subtract

Count back to subtract. Write the difference.
You can use the number line to help.

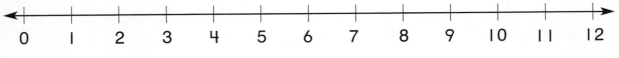

1. 8 − 2
2. 9 − 1
3. 11 − 3
4. 7 − 1
5. 6 − 2
6. 11 − 2

7. 12 − 3
8. 10 − 1
9. 9 − 2
10. 8 − 3
11. 7 − 3
12. 6 − 1

13. 10 − 2
14. 6 − 3
15. 8 − 1
16. 7 − 2
17. 10 − 3
18. 9 − 3

Related Addition and Subtraction Facts to 12

Write each sum or difference.
Circle the related facts in each row.

19. 12 − 4 = ___
20. 12 − 6 = ___
21. 8 + 4 = ___

22. 5 + 7 = ___
23. 11 − 7 = ___
24. 12 − 5 = ___

25. 9 − 6 = ___
26. 9 − 3 = ___
27. 5 + 3 = ___

Name _____

Use a Number Line to Count Back

Vocabulary
count back

Learn

You can use the number line to help you count back.

Find the difference for 12 − 3.

Start at 12.
Count back 3 spaces.
11, 10, 9

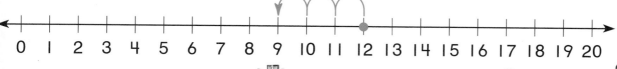

12 − 3 = __9__

Check

Count back to subtract. Write the difference.
Use the number line to help.

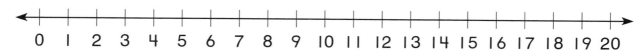

1. 9 − 2 = ____

2. 10 − 1 = ____

3. 8 − 3 = ____

4. 11 − 3 = ____

5. 7 − 1 = ____

6. 6 − 3 = ____

7. 8 − 2 = ____

8. 11 − 2 = ____

9. 10 − 3 = ____

10. 9 − 1 = ____

Explain It • Daily Reasoning

Pat drew this number line to find 10 − 2 and got 12. What mistake did she make?

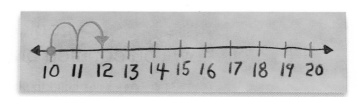

Chapter 19 • Subtraction Facts and Strategies   three hundred seventeen **317**

Practice and Problem Solving

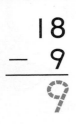
18
− 9
___
9

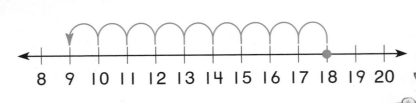

Use the number line to subtract.

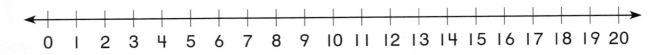

1. 20
   −10
   ___

2. 13
   − 6
   ___

3. 17
   − 9
   ___

4. 14
   − 5
   ___

5. 11
   − 3
   ___

6. 16
   − 9
   ___

7. 15
   − 8
   ___

8. 18
   − 9
   ___

9. 11
   − 2
   ___

10. 12
   − 4
   ___

11. 14
   − 6
   ___

12. 13
   − 5
   ___

13. 16
   − 7
   ___

14. 10
   − 2
   ___

15. 12
   − 3
   ___

16. 16
   − 8
   ___

17. 15
   − 9
   ___

18. 17
   − 8
   ___

Problem Solving
Visual Thinking

19. Write the number sentence that tells about the number line.

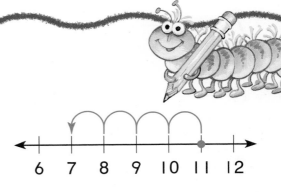

 Write About It • Look at Exercise 19.
Explain how you counted on the number line.

🏠 **HOME ACTIVITY** • Ask your child to show how to subtract 18 − 9 on the number line.

Name _____

Doubles Fact Families

Vocabulary
fact family

Learn

These facts use the same two numbers.
Together, they make a doubles fact family.

4 + 4 = __8__   __8__ ◯ __4__ ◯ __4__

Check

Write the sum for the doubles addition fact.
Write the subtraction fact that is in the same family.

1. 6 + 6 = ____   ____ ◯ ____ ◯ ____

2. 3 + 3 = ____   ____ ◯ ____ ◯ ____

3. 8 + 8 = ____   ____ ◯ ____ ◯ ____

4. 5 + 5 = ____   ____ ◯ ____ ◯ ____

5. 9 + 9 = ____   ____ ◯ ____ ◯ ____

6. 7 + 7 = ____   ____ ◯ ____ ◯ ____

Explain It • Daily Reasoning

Why are there only two facts in doubles fact families?

Chapter 19 • Subtraction Facts and Strategies

three hundred nineteen **319**

Name _____

Practice the Facts

Learn

There are many ways to find sums and differences!

I can count on, make a ten, or use doubles or doubles plus one to add.

I can count back or use a related fact to subtract.

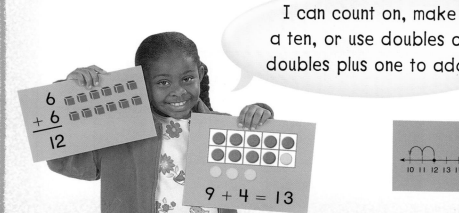

Check

Add or subtract.

1. 9 +5 = 14
2. 12 − 3
3. 7 +9
4. 13 − 8
5. 14 − 5
6. 9 +9

7. 5 +8
8. 11 − 4
9. 6 +6
10. 10 − 4
11. 13 − 6
12. 4 +7

13. 8 +4
14. 14 − 7
15. 5 +7
16. 18 − 9
17. 18 − 8
18. 10 + 9

Explain It • Daily Reasoning

What ways could you use to find the sum for 7 + 6? What ways could you use to find the difference for 12 − 3?

Chapter 20 • Addition and Subtraction Practice

Practice and Problem Solving

1. Solve the number puzzle.
   Write each sum or difference.
   The problems go across and down.

	16	−	8	=	8		15	−	7	=	
−	9			+	2					+	5
	7					+	8	=			
								−	9		
	10	+	5	=			17				
+	8					−	8				
		−	9	=				+	7	=	

Problem Solving
Logical Reasoning

2. The sum for two of these numbers is 14.
   The difference for the same two numbers is 2.
   What are the two numbers?

5	8
6	9

   _____ and _____

 Write About It • Look at Exercise 2.
The sum for two other numbers is also 14.
What are the two numbers?
What is the difference for those numbers?

HOME ACTIVITY • With your child, make flash cards for the addition facts with sums of 10 through 20. Ask your child to choose a card, say the sum, and then tell you a related subtraction fact. (For example: 8 + 7 = 15, 15 − 7 = 8)

Fact Families to 20

Vocabulary: fact family

Learn

Fact Family
9 + 8 = 17
So, 8 + 9 = 17
17 − 8 = 9
17 − 9 = 8

You can use one fact in a *fact family* to help you write the other facts in the same family.

Check

Write the sum or difference.
Circle the two facts if they are in the same fact family.

1.
   6 + 7 = __13__
   13 − 6 = __7__

2.
   18 + 2 = ____
   20 − 10 = ____

3.
   9 + 3 = ____
   12 − 9 = ____

4.
   10 + 9 = ____
   10 − 3 = ____

5.
   15 − 7 = ____
   7 + 8 = ____

6.
   5 + 9 = ____
   14 − 5 = ____

7.
   18 − 9 = ____
   9 + 9 = ____

8.
   19 − 9 = ____
   9 + 2 = ____

Explain It • Daily Reasoning

Which facts are in the same family as 13 − 9 = 4?
How do you know?

Practice and Problem Solving

Write the sum or difference.
Color all the facts in the same fact family to match.

1. 9 + 7 = 16
2. 6 + 8 = ___
3. 17 − 9 = ___
4. 6 + 9 = ___
5. 15 − 6 = ___
6. 16 − 9 = ___
7. 8 + 6 = ___
8. 17 − 8 = ___
9. 8 + 9 = ___
10. 9 + 6 = ___
11. 16 − 7 = ___
12. 14 − 6 = ___
13. 14 − 8 = ___
14. 9 + 8 = ___
15. 15 − 9 = ___
16. 7 + 9 = ___

Problem Solving
Application

17. Write the number sentence that is missing from this fact family.

____ ◯ ____ ◯ ____

$5 + 7 = 12$
$12 − 7 = 5$
$7 + 5 = 12$

 Write About It • Look at Exercise 17. Explain how you figured out the missing number sentence.

⬡ **HOME ACTIVITY** • Say an addition or subtraction fact to your child. Ask him or her to tell another fact that is in the same fact family. (For example: 7 + 6 = 13, 13 − 6 = 7)

Name _____

Algebra: Ways to Make Numbers to 20

Explore

You can make the number 19 in different ways.

4 + 9 + 6

20 − 1

15 + 4

You can add or subtract to make 19.

Connect

Use 🎲🎲🎲. 
Circle all the ways to make the number at the top.

1.
18
(9 + 9)
8 + 4 + 4
19 − 1
5 + 5 + 7
20 − 8
15 + 2
20 − 2

2.
20
13 + 6 + 1
5 + 4 + 10
20 − 0
14 + 6
5 + 7 + 8
12 + 7
2 + 8 + 7

Explain It • Daily Reasoning

Look at Exercise 2. What are three other ways to make 20?

Practice and Problem Solving

Use ▢▢▢.
Circle all the ways to make the number at the top.

1.
15
(6 + 4 + 5)
8 + 2 + 5
10 + 5
7 + 9
15 − 0

2.
17
6 + 10
17 − 0
5 + 5 + 7
18 − 0
7 + 10

3.
14
4 + 10
19 − 9
2 + 3 + 9
14 − 0
7 + 5 + 1

4.
16
6 + 4 + 6
10 + 6
17 + 1
8 + 8
7 + 3 + 5

Problem Solving
Application

5. Circle ways to show 12.

   6 + 7   twelve   ❄❄❄❄ ❄❄❄❄ ❄❄❄❄

 Write About It • Use pictures, words, and numbers to show 13.

 HOME ACTIVITY • Ask your child to tell you three ways to make 20.

Name _____

Problem Solving Strategy
Make a Model

10 girls are sledding.
7 more come.
(How many girls are sledding?)

UNDERSTAND

What do you need to find out?
Circle the question.

PLAN

How will you solve this problem?
You can make a model.
Use and ▪ to show the groups of girls.

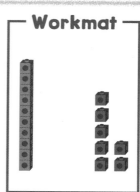
Workmat

SOLVE

I have 10. I can count 7 more.

There are ___17___ girls.

CHECK

Does your answer make sense? Explain.

Use Workmat 1, ▪, and ▪.
Draw the ▪ and ▪ you use.
Write the answer.

THINK:
I have 17. How many more do I count to get to 20?

1. Tom's class has 20 children. 17 children are here today. How many children are absent?

   _____ children

Chapter 20 • Addition and Subtraction Practice

three hundred thirty-seven **337**

Problem Solving Practice

Use Workmat 1, 🎲, and 🎲.
Draw the 🎲 and 🎲 you use.
Write the answer.

Keep in Mind!
Understand
Plan
Solve
Check

1. 10 boys are skating.
   There are 18 boys in all.
   How many boys are not skating?

   _____ boys

2. Robin sees 9 children.
   9 more are hiding.
   How many children are there in all?

   _____ children

3. Kate makes 8 snowballs.
   Then she makes 7 more.
   How many snowballs does she have now?

   _____ snowballs

4. 15 mittens are missing.
   Jan finds 5 mittens.
   How many mittens are still missing?

   _____ mittens

HOME ACTIVITY • Ask your child how he or she decided to solve each problem.

Name _____

Extra Practice

Add or subtract.

1. 7
   +8
   ___

2. 11
   − 4
   ___

3. 6
   +6
   ___

4. 16
   − 7
   ___

5. 13
   − 8
   ___

6. 5
   +9
   ___

Write the sum or difference.
Circle the two facts if they are in the same fact family.

7. 10 + 0 = _____

   17 − 8 = _____

8. 7 + 9 = _____

   16 − 7 = _____

---

9. Use 🎲🎲🎲.
   Circle all the ways
   to make the number
   at the top.

20
10 + 10
9 + 9
7 + 3 + 10
12 − 3
3 + 4 + 3

Problem Solving

Use Workmat 1, 🎲, and 🎲.
Draw the 🎲 and 🎲 you use.
Write the answer.

10. Ms. Lee sees 8 children. 7 more
    are hiding. How many children
    are there in all?

    _____ children

Name _____

✅ Review/Test

Concepts and Skills

Add or subtract.

1. 6   2. 15   3. 9   4. 14   5. 17   6. 10
   +7   − 5   +9   − 6   − 9   + 8
   ──   ──   ──   ──   ──   ──

Write the sum or difference.
Circle the pair of facts if they are in the same fact family.

7. 8 + 7 = _____   8. 19 − 9 = _____

    15 − 8 = _____   9 + 9 = _____

9. Use 🎲🎲🎲.
   Circle all the ways
   to make the number
   at the top.

11
6 + 5
11 − 0
4 + 2 + 6
8 + 3
5 + 5 + 1

Problem Solving

Use Workmat 1, 🎲, and 🎲. Draw the
🎲 and 🎲 you use. Write the answer.

10. There are 15 children marching.
    Some children leave. 9 children
    are still marching. How many
    children left?

    _____ children

340 three hundred forty

Name _____

★Standardized Test Prep
Chapters 1–20

Choose the answer for questions 1–5.

1. Which object is shaped most like a sphere?

○ ○ ○ ○

2. What is the difference?

   $18 - 9 = $ _____

6	7	8	9
○	○	○	○

3. What is the sum?

   $10 + 6 = $ _____

13	15	16	20
○	○	○	○

4. Which fact is in the same family as $8 + 9 = 17$?

$17 - 8 = 9$	$17 - 7 = 10$	$9 - 8 = 1$	$1 + 8 = 9$
○	○	○	○

5. Which is a way to make 18?

$9 + 8$	$9 + 4 + 5$	$10 + 7$	$10 - 6$
○	○	○	○

Show What You Know

6. Solve. Draw 🎲 to explain.
   Write a number sentence.

   14 children play in the snow.
   6 children are still playing.
   How many children went home?

   children

IT'S IN THE BAG
The Hungry Prince's Crown

PROJECT You will make a crown with your hardest math facts.

You Will Need

- Paper plate
- Pattern tracer
- Crayons
- Scissors

Directions

1. Put the paper plate upside down in front of you. Write your hardest math facts around the outside. Fold the paper plate in half.

2. Place the pattern on top of your plate. Trace the cut lines on your plate.

3. Cut on the lines you traced.

4. Open the plate. Fold back the points to make the top of the crown. Decorate your crown.

The Hungry Prince

written by Lucy Floyd
illustrated by Alexi Natchev

◆ This book will help me review doubles plus one.

This book belongs to _____.

Once there was a very hungry prince.
"I have only 5 muffins," said the prince.
"I need MORE!"

The cook gave him 5 more.

"Now I have 5 + 5 = ____ muffins,"
said the hungry prince. "I still need more."

The cook gave him 1 more muffin. "Goody!" said the hungry prince.

"Now I have 5 + 6 = ____ muffins!"

He ate every one of them.

"I am still hungry," said the prince.
The cook gave him 6 rolls.
"I need MORE!" said the prince.

The cook gave him 6 more rolls.

"Now I have 6 + 6 = ____ rolls!" said the hungry prince. "I still need more."

The cook gave him 1 more roll.
"Goody!" said the hungry prince.

"Now I have 6 + 7 = _____ rolls!"

He ate every one of them.

"I am still hungry," said the prince.
The cook gave him 7 bagels.
"I need MORE!" said the prince.

The cook gave him 7 more bagels.

"Now I have 7 + 7 = ____ bagels!"
said the hungry prince. "I still need more."

The cook gave him 1 more bagel. "Goody!" said the hungry prince.

"Now I have 7 + 8 = ____ bagels! Should I eat them all?"

WHAT DID THE PRINCE DO?

The prince did eat them all!

Then he was a sick prince, but he was NOT a hungry prince any more!

PROBLEM SOLVING ON LOCATION

At the Park

Each year, more than 20 million people visit Central Park in New York City.

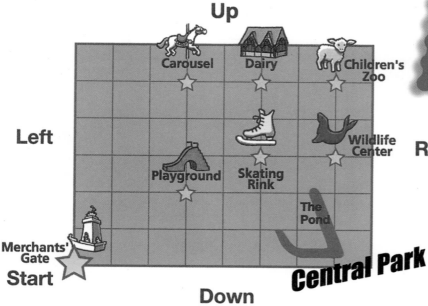

Visit some places in Central Park.
Fill in the blanks to show your path.

1. Go _____. Go _____.

   Where are you? _____

2. Go _____. Go _____.

   Where are you? _____

3. Go _____. Go _____.

   Where are you? _____

Unit 4 • Chapters 15–20   three hundred forty-three **343**

Name _____

CHALLENGE

Repeated Addition

Each bike has 2 wheels.
How many wheels are there in all?

__2__ + __2__ + __2__ + __2__ = __8__

You can add to find how many wheels there are.

Complete the number sentence.

1. Each boat has 2 sails.
   How many sails are there in all?

   ____ + ____ + ____ + ____ + ____ = ____

2. Each car has 4 wheels.
   How many wheels are there in all?

   ____ + ____ + ____ + ____ + ____ = ____

3. Each swing set has 3 swings.
   How many swings are there in all?

   ____ + ____ + ____ + ____ = ____

CHALLENGE

Name _____

✓ Study Guide and Review

Vocabulary

Use 🖍 to color the **circles**.
Use 🖍 to color the **triangles**.
Use 🖍 to color the **rectangles** and **squares**.

1.

Skills and Concepts

Color each solid that will stack.

2.

3. Write how many sides and vertices.

_____ sides

_____ vertices

4. Draw a line of symmetry to show two matching parts.

Write the sum.

5. 9
   $+5$
   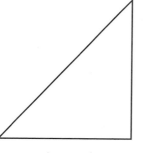

6. 8
   $+4$
   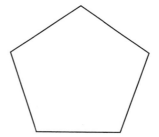

Add.

7. $9 \quad\quad 10$
   $+6 \quad\quad +5$

8. $7 \quad\quad 10$
   $+4 \quad\quad +1$

9. $8 \quad\quad 10$
   $+5 \quad\quad +3$

Unit 4 • Study Guide and Review

Subtract.

10. 15 − 7

11. 18 − 9

12. 20 − 10

13. 13 − 5

14. 17 − 8

15. 16 − 8

Write the sum for the doubles addition fact.
Write the subtraction fact that is in the same family.

16. 9 + 9 = ____   ____ ◯ ____ ◯ ____

17. 7 + 7 = ____   ____ ◯ ____ ◯ ____

18. 8 + 8 = ____   ____ ◯ ____ ◯ ____

Problem Solving

Find the pattern. Circle the mistake.
Draw the correct shape.

19.

20.

Name _____

✓ Performance Assessment

How to Make a House

Sal had these blocks.

| 10  | 7  | 10  | 8  |

- He used 16 blocks to build a house.
- All the blocks he used had faces that were squares or triangles.

Draw 16 blocks Sal could have used. Write the number sentence to show the blocks he used.

Show your work.

Unit 4 • Performance Assessment

Name _____

TECHNOLOGY

The Learning Site • Addition Surprise

1. Go to www.harcourtschool.com.
2. Click on 🐻.
3. Drag the first number tile to start.

Practice and Problem Solving

Use 🎲 🎲 🎲.
Circle all the ways to make the number at the top.

1.
16
8 + 9
4 + 14
8 + 8
14 + 2 + 2
11 + 5

2.
20
11 + 9
10 + 10
12 + 7
7 + 12 + 1
8 + 5 + 6

Write the sums. Circle the pair of facts if they are in the same fact family.

3. 9 + 6 = _____

   10 + 5 = _____

4. 6 + 7 = _____

   7 + 6 = _____

348 three hundred forty-eight

Dear Family,

In Unit 4 we learned about geometry and about addition and subtraction to 20. Here is a game for us to play together. This game will give me a chance to share what I have learned.

Love,

_____

Directions
1. Put a game piece at START. Cover each button with a coin.
2. Your partner gives you directions such as, "Go down 2. Go left 1. Where are you?"
3. Move the game piece. Give your partner the coin where you land.
4. Take turns.
5. Play until all of the coins are taken.
6. The player with more coins wins.

Materials
- 10 pennies
- 10 nickels
- 10 dimes
- 2 game pieces or beans

Find the Pattern

START

Dear Family,

During the next few weeks, we will learn about fractions, money, and time. Here is important math vocabulary and a list of books to share.

Love,

_____

Vocabulary

$\frac{1}{2}$ one half

$\frac{1}{3}$ one third

$\frac{1}{4}$ one fourth

minute hand

hour hand

Vocabulary Power

Two equal parts are halves.

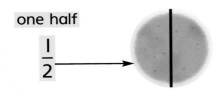

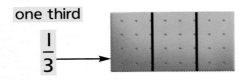

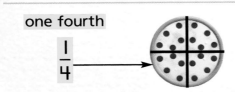

BOOKS TO SHARE

To read about fractions, money, and time with your child, look for these books in your library.

Eating Fractions,
by Bruce McMillan,
Scholastic, 1991.

Fraction Action,
by Loreen Leedy,
Holiday House, 1996.

26 Letters and 99 Cents,
by Tana Hoban, William Morrow, 1995.

Isn't It Time?
by Judy Hindley,
Candlewick, 1996.

 Visit *The Learning Site* for additional ideas and activities. www.harcourtschool.com

CHAPTER 21

Fractions

FUN FACTS

On an average size pizza, $\frac{1}{2}$ of the pizza's weight is crust and $\frac{1}{4}$ of the weight is cheese.

Theme: At the Pizza Party

Name _____

✓ Check What You Know

Equal Parts

Circle the shape that is divided into 2 equal parts.

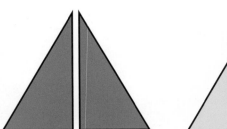 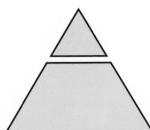 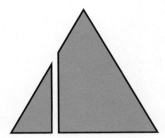

Circle the shape that is divided into 3 equal parts.

 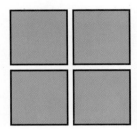 

Circle the shape that is divided into 4 equal parts.

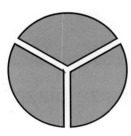 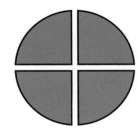 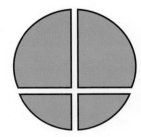

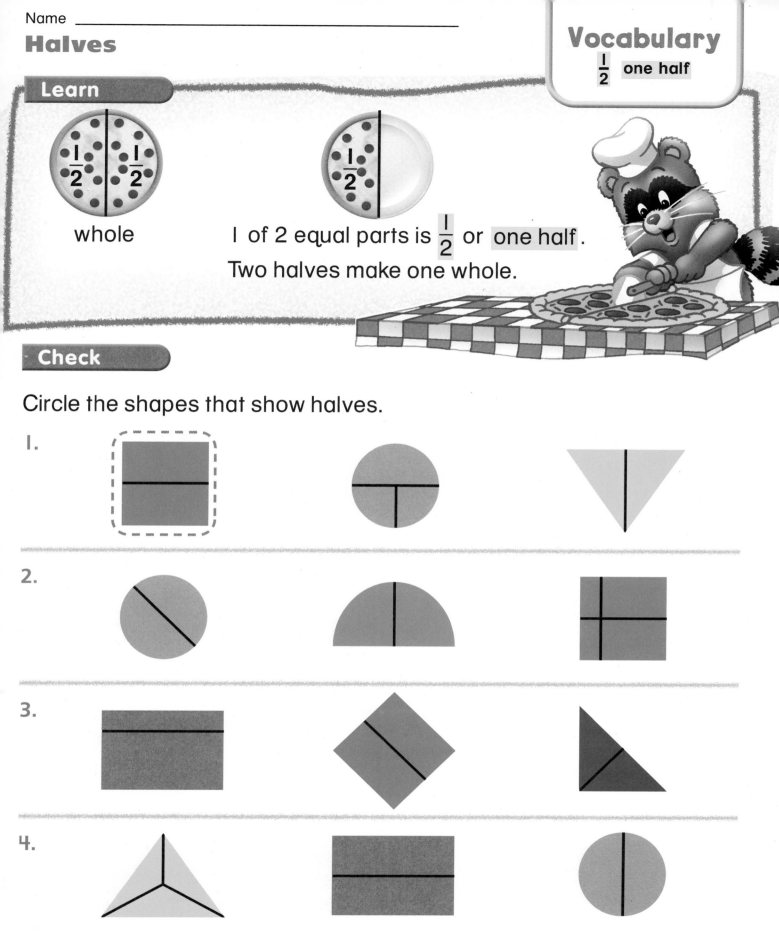

Practice and Problem Solving

Find the shapes that show halves.
Color $\frac{1}{2}$.

This has 2 equal parts.

1.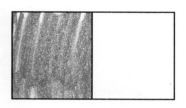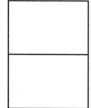

2.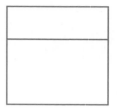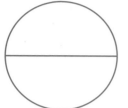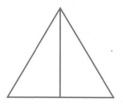

3.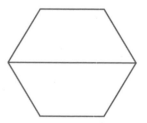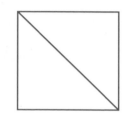

4.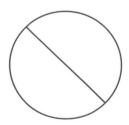

Problem Solving
Application

5. Draw a line on each cracker to show different ways to make halves.

Write About It • Write a story about sharing a sandwich with a friend. Use **one half** in your story.

HOME ACTIVITY • Give your child three sheets of paper, each a different size. Ask him or her to fold each sheet in half and to name each part as one half.

Name _____

⭐Standardized Test Prep
Chapters 1-21

Choose the answer for questions 1-5.

1. Which shape shows halves?

2. Which shape has $\frac{1}{3}$ colored in?

3. Which shape shows fourths?

4. What fraction does the colored part show?

$\frac{1}{4}$ $\frac{1}{2}$ $\frac{1}{3}$
○   ○   ○

5. Which is a way to make 20?

10 + 10   10 + 4 + 4   10 + 6   10 + 4
  ○   ○   ○   ○

Show What You Know

6. Explain how the children share the pizza. Draw a picture that matches the clues.

Ellen and Mara have a pizza. They share equal parts.

Color $\frac{1}{2}$ of the pizza.

Chapter 21   three hundred sixty-three **363**

Name _____

MATH GAME

Pizza Party

Play with a partner.

1. Put your 🨠 on START.
2. Spin the 🌀.
3. If the parts in that circle are not equal, your turn is over.
4. If the parts are equal, count them.
5. Move your 🨠 that many spaces. If you land on a space with equal parts, spin again.
6. The first player to get to the party wins.

You will need

2 🨠

CHAPTER 22

Counting Pennies, Nickels, and Dimes

FUN FACTS

The face of a penny can hold about 30 drops of water.

Theme: It's in the Bank

Name _____

✓ Check What You Know

Penny

Count the pennies.
Write how many cents.

1.   _____ ¢

2.
        _____ ¢

Nickel

Write how many cents.

3. _____ ¢

4.   _____ ¢

5.    _____ ¢

Dime

Write how many cents.

6.   _____ ¢

7.   _____ ¢

Name _____

Pennies and Nickels

Explore

A penny is worth 1 cent.
A nickel is worth 5 cents.

Vocabulary
penny
nickel

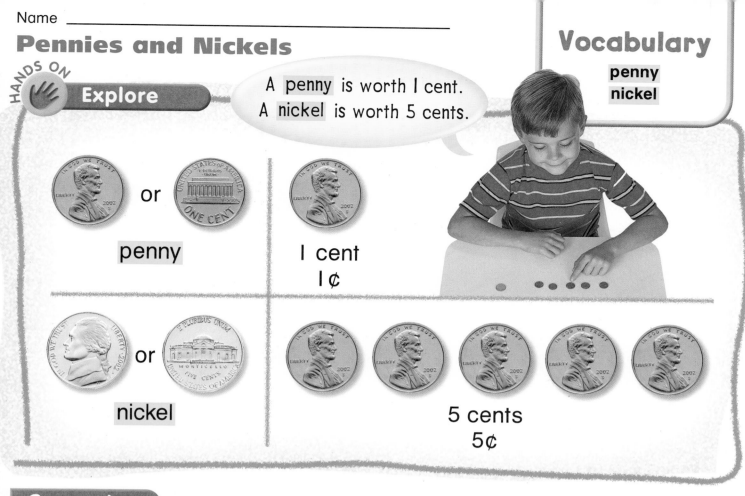

Connect

Use and . Draw and label them.
Count by ones or fives. Write the amount.

1. 3 pennies   3 ¢

2. 4 pennies   ☐ ¢

3. 2 nickels   ☐ ¢

4. 3 nickels   ☐ ¢

Explain It • Daily Reasoning

Which is worth more, two pennies
or one nickel? Explain.

Chapter 22 • Counting Pennies, Nickels, and Dimes   three hundred sixty-seven **367**

Practice and Problem Solving

Count by ones or fives. Write the amount.

1.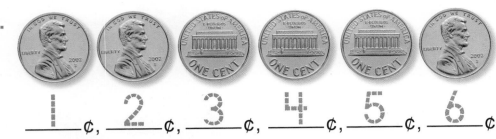

   1 ¢, 2 ¢, 3 ¢, 4 ¢, 5 ¢, 6 ¢   6 ¢

2.

   ___ ¢, ___ ¢, ___ ¢, ___ ¢, ___ ¢, ___ ¢, ___ ¢   ☐ ¢

3.

   ___ ¢, ___ ¢, ___ ¢, ___ ¢   ☐ ¢

4.

   ___ ¢, ___ ¢, ___ ¢, ___ ¢, ___ ¢   ☐ ¢

Problem Solving

Logical Reasoning

5. Mary has 2 nickels. Sam has the same amount of money in pennies. How many pennies does Sam have? _____ pennies

Write About It • Look at Exercise 5. What would Mary say to count her money? Explain.

 HOME ACTIVITY • Set out a group of pennies or nickels. Have your child count by ones or fives to find the value of the group of coins.

Name _____

Pennies and Dimes

Vocabulary
dime

Explore

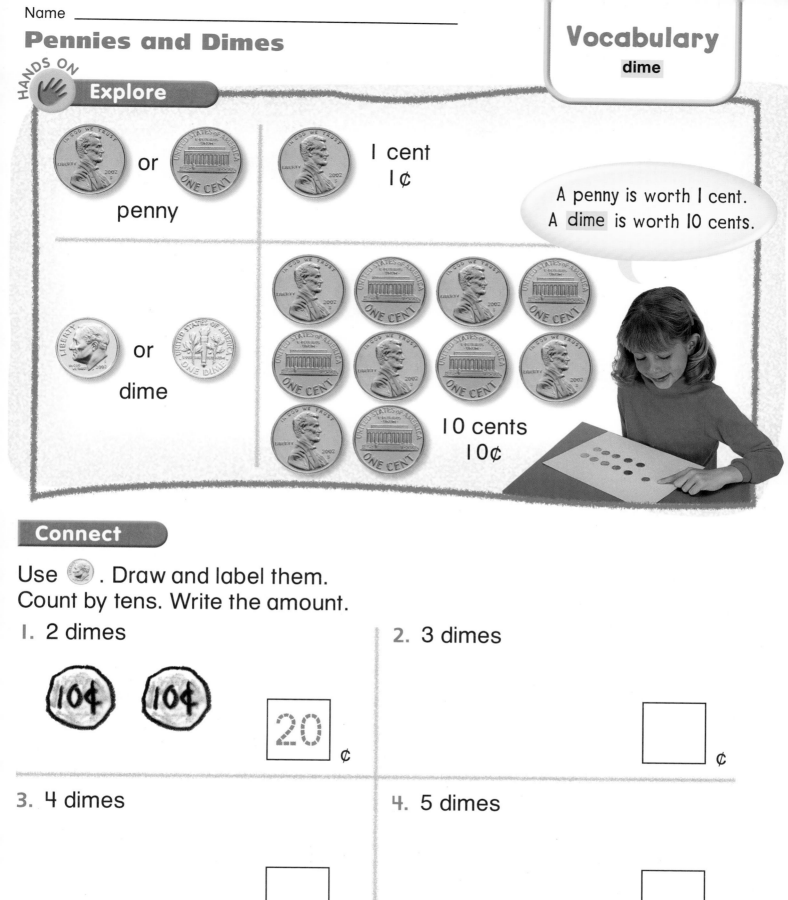

A penny is worth 1 cent.
A dime is worth 10 cents.

Connect

Use 🪙. Draw and label them.
Count by tens. Write the amount.

1. 2 dimes

   (10¢) (10¢)   [20] ¢

2. 3 dimes

   [ ] ¢

3. 4 dimes

   [ ] ¢

4. 5 dimes

   [ ] ¢

Explain It • Daily Reasoning

How many pennies equal two dimes?
Explain how you know.

Practice and Problem Solving

Count by tens. Write the amount.

1.

   __10__ ¢, __20__ ¢, __30__ ¢

   __30__ ¢

2.

   _____ ¢, _____ ¢, _____ ¢, _____ ¢, _____ ¢

    ¢

3.

   _____ ¢, _____ ¢, _____ ¢, _____ ¢

    ¢

4.

   _____ ¢, _____ ¢, _____ ¢, _____ ¢, _____ ¢, _____ ¢ ☐ ¢

Problem Solving
Mental Math

5. Joanne has 9 dimes. How much money does she have? ¢

6. Paul has 7 dimes. How much money does he have? ¢

 Write About It • How much money would you have if you had 8 dimes? Explain.

🏠 **HOME ACTIVITY** • Have your child count groups of dimes by tens and tell the value of each group.

Name _____

Count Groups of Coins

Learn

Count by tens.   Then count on by ones.

__10__¢, __20__¢, __30__¢, __40__¢, __41__¢, __42__¢, __43__¢   __43__¢

Check

Count by tens. Then count on by ones.
Write the amount.

1.

   ____¢, ____¢, ____¢, ____¢, ____¢   ☐ ¢

2.

   ____¢, ____¢, ____¢, ____¢, ____¢, ____¢   ☐ ¢

3.

   ____¢, ____¢, ____¢, ____¢, ____¢, ____¢   ☐ ¢

Explain It • Daily Reasoning

You want to count this group of coins.
With which coin would you start? Why?

Chapter 22 • Counting Pennies, Nickels, and Dimes   three hundred seventy-one **371**

Practice and Problem Solving

Count by fives.   Then count on by ones.

__5__ ¢, __10__ ¢, __15__ ¢, __20__ ¢, __25__ ¢, __26__ ¢, __27__ ¢   __27__ ¢

Count. Write the amount.

1.

___ ¢, ___ ¢, ___ ¢, ___ ¢, ___ ¢, ___ ¢, ___ ¢   ☐ ¢

2.

___ ¢, ___ ¢, ___ ¢, ___ ¢, ___ ¢, ___ ¢, ___ ¢   ☐ ¢

Problem Solving
Application

3. Write the number. Draw and label dimes and pennies to show the same amount in cents. Write the amount.

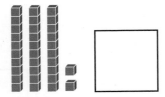 ☐

☐ ¢

Write About It • Look at Exercise 3. How many pennies would it take to show the same amount? Explain.

 HOME ACTIVITY • Have your child count groups of dimes and pennies or groups of nickels and pennies and tell the value of each group.

Name _____

Count Collections

Learn

Count by tens. Count by fives. Then count on by ones.

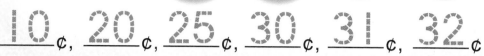

10 ¢, 20 ¢, 25 ¢, 30 ¢, 31 ¢, 32 ¢   [32] ¢

Check

Count. Write the amount.

1.

____ ¢, ____ ¢, ____ ¢, ____ ¢, ____ ¢, ____ ¢   ☐ ¢

2.

____ ¢, ____ ¢, ____ ¢, ____ ¢, ____ ¢, ____ ¢   ☐ ¢

3.

____ ¢, ____ ¢, ____ ¢, ____ ¢, ____ ¢, ____ ¢   ☐ ¢

Explain It • Daily Reasoning

What are three ways to show 16¢?
Use coins to prove your answer.

Chapter 22 • Counting Pennies, Nickels, and Dimes

Practice and Problem Solving

Start counting with the coins that have the greatest value.

Count. Write the amount.

1.    36 ¢

2.   ___ ¢

3.     ___ ¢

4.    ___ ¢

5.  ___ ¢

6.    ___ ¢

Problem Solving
Logical Reasoning

Draw and label the coins.

7. You have 72¢.
   You have 9 coins.
   What coins do you have?

 Write About It • Look at Exercise 7. How many pennies would it take to show the same amount? Explain.

🏠 **HOME ACTIVITY** • Have your child count groups of dimes, nickels, and pennies. Ask your child to tell you the value of each group.

374 three hundred seventy-four

Problem Solving Strategy
Make a List

Earl wants to buy a yo-yo for 15¢.

(In what ways can he use and 🪙 to make 15¢?)

UNDERSTAND

What do you need to find out?
Circle it.

PLAN

How can you solve the problem?
Make a list of the coins Earl can use.

SOLVE

Draw and label the coins that will make 15¢.

dimes	nickels
0	5¢ 5¢ 5¢

CHECK

Does each way equal 15¢?
Explain.

Chapter 22 • Counting Pennies, Nickels, and Dimes

Problem Solving Practice

Keep in Mind!
Understand
Plan
Solve
Check

Tina wants to buy a toy for 20¢. In what ways can she use 🪙, 🪙, and 🪙 to make 20¢?

List six ways to make 20¢.
Use 🪙, 🪙, and 🪙.
Draw and label the coins.

Ways to Make 20¢		
dimes	nickels	pennies
10¢ 10¢	0	0
		0
0		

HOME ACTIVITY • Place pennies, nickels, and dimes on the table. Ask your child to show all the ways to make 20¢.

Name _____

Extra Practice

Count. Write the amount.

1.

____¢, ____¢, ____¢, ____¢, ____¢   ☐ ¢

2.

____¢, ____¢, ____¢, ____¢   ☐ ¢

3.

____¢, ____¢, ____¢   ☐ ¢

4.

____¢, ____¢, ____¢, ____¢, ____¢   ☐ ¢

Problem Solving

5. List two ways to make 25¢.
   Use 🪙 and 🪙.
   Draw and label the coins.

dimes	nickels

Chapter 22 • Counting Pennies, Nickels, and Dimes

Name _____

✓ Review/Test

Concepts and Skills
Count. Write the amount.

1.

   ____¢, ____¢, ____¢, ____¢, ____¢, ____¢   ☐ ¢

2.

   ____¢, ____¢, ____¢, ____¢, ____¢, ____¢, ____¢   ☐ ¢

3.

   ____¢, ____¢, ____¢, ____¢, ____¢   ☐ ¢

4.

   ____¢, ____¢, ____¢, ____¢, ____¢, ____¢, ____¢   ☐ ¢

Problem Solving

5. List two ways to make 30¢.
   Use 🪙, 🪙, and 🪙.
   Draw and label the coins.

dimes	nickels	pennies

378 three hundred seventy-eight

Name _____

★Standardized Test Prep
Chapters 1–22

Choose the answer for questions 1–5.

1. Find the pattern. Which should come next?

2. How much money is shown?

   3¢   5¢   15¢   20¢
   ○   ○   ○   ○

3. How much money is shown?

   4¢   6¢   10¢   12¢
   ○   ○   ○   ○

4. How much money is shown?

   10¢   20¢   40¢   50¢
   ○   ○   ○   ○

5. Which is the total amount?

   13¢   15¢   17¢   20¢
   ○   ○   ○   ○

Show What You Know

6. Theo wants to buy a jump rope for 25¢. Which ways can he use 🪙 and 🪙 to make 25¢?

   Explain different ways to make 25¢. Use 🪙 and 🪙 to list the ways. Draw to show what you choose.

Ways to Make 25¢	
dimes	nickels

Chapter 22   three hundred seventy-nine **379**

Name _____

Half Dollar and Dollar

Vocabulary
half dollar
dollar

Hands On Explore

or

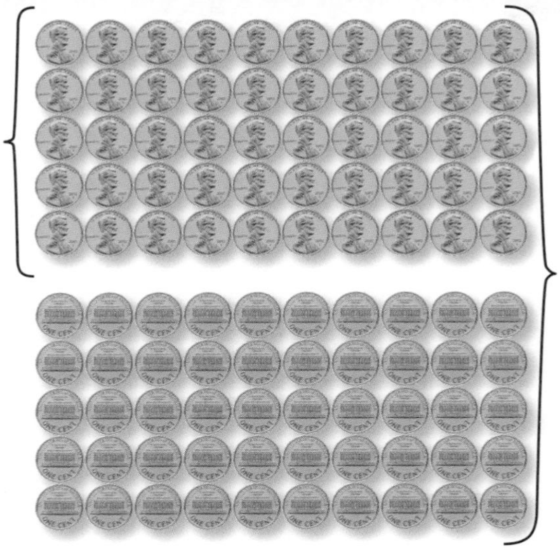

1 half dollar = 50¢

1 dollar = 100¢

Connect

Draw and label the coins. Write how many.

1. Show how many quarters equal 1 dollar.

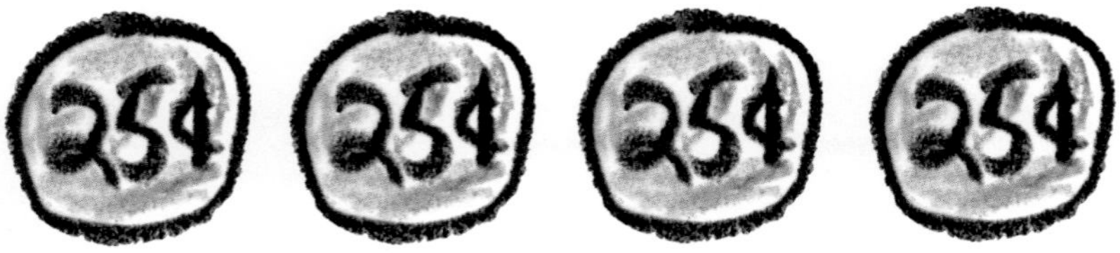

___4___ quarters = 1 dollar

2. Show how many dimes equal 1 half dollar.

_____ dimes = 1 half dollar

Explain It • Daily Reasoning

Explain how you could find out how many nickels equal 1 dollar.

Chapter 23 • Using Money

three hundred eighty-seven **387**

Practice and Problem Solving

THINK:
1 dollar = 100¢
1 half dollar = 50¢

Draw and label the coins. Write how many.

1. Show how many dimes equal 1 dollar.

   _____ dimes = 1 dollar

2. Show how many quarters equal 1 half dollar.

   _____ quarters = 1 half dollar

3. Show how many nickels equal 1 half dollar.

   _____ nickels = 1 half dollar

Problem Solving
Mental Math

Solve. Write the amount.

4. Kevin saved 4 dimes in one week. In the next week, he saved 1 half dollar. How much money did he have then?

   _____ ¢

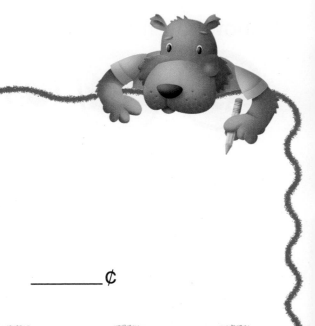

 Write About It • Look at Exercise 4. Kevin saves one dime this week. Draw a picture to show how much money Kevin has now.

HOME ACTIVITY • Show your child a dollar bill. Ask him or her to show the same amount of money, using quarters, dimes, and nickels.

Name _____

Compare Values

Learn

Write the value for each group.
Which amount is greater? Circle it.

(50) ¢   35 ¢

Check

Write the value for each group.
Circle the amount that is greater.

1.

_____ ¢   _____ ¢

2.

_____ ¢   _____ ¢

Explain It • Daily Reasoning

Compare the values of 1 dollar, 1 quarter, and 1 half dollar.
Put them in order from least value to greatest value.

Practice and Problem Solving

Write the amount for each group.
Circle the amount that is greater.

1.  _____ ¢    _____ ¢

2.  _____ ¢   _____ ¢

3.  _____ ¢    _____ ¢

Problem Solving
Application

4. Cary wants quarters.
   He has 6 dimes and 3 nickels.
   Draw quarters to show the
   same amount.

 Write About It • Cary gets one more coin.
Now he has 1 dollar. Draw and label the coin.
Tell how you know.

HOME ACTIVITY • Show your child two groups of coins, each worth one dollar or less. Have your child tell you the amount for each group. Then ask which amount is greater.

Name _____

Same Amounts

HANDS ON Explore

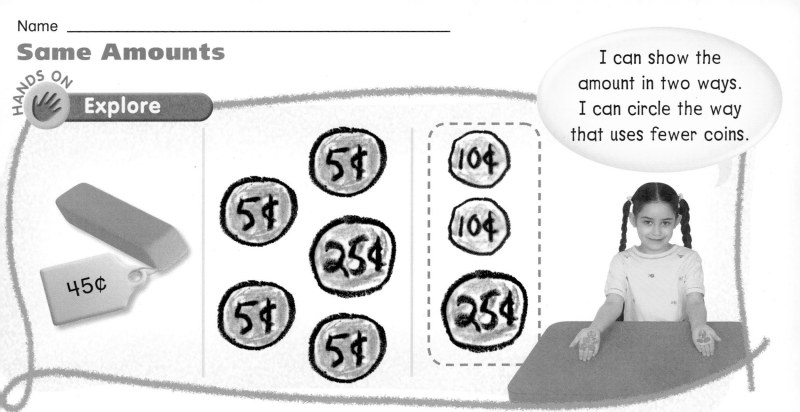

I can show the amount in two ways. I can circle the way that uses fewer coins.

Connect

Use coins. Show the amount in two ways.
Draw and label the coins.
Circle the way that uses fewer coins.

1.

2.

Explain It • Daily Reasoning

How could you show this amount using fewer coins? Is there more than one way? Explain.

Chapter 23 • Using Money

three hundred ninety-one **391**

Practice and Problem Solving

Use coins. Show the amount in two ways.
Draw the coins.
Circle the way that uses fewer coins.

1.

2.

3.

Problem Solving
Visual Thinking

Draw another way to show the same amount.

4. Luis shows the same amount in two ways. One way has 1 quarter and 2 dimes. One way has 9 nickels.

 Write About It • Look at Exercise 4. Use words to tell which group uses the fewest coins.

HOME ACTIVITY • Name an amount of money that is less than 50¢. Have your child use pennies, nickels, and dimes to show the amount in different ways. Then have your child point to the way that uses the fewest coins.

Problem Solving Strategy
Act It Out

You want to buy these two things. What coins could you use?

UNDERSTAND

What do you want to find out?
Circle it.

PLAN

How will you solve the problem?
Act it out. Use coins.

SOLVE

Show the coins you would use.
Draw and label the coins.

CHECK

Does your answer make sense?
Explain.

Show the coins you would use.
Draw and label the coins.

1.

Chapter 23 • Using Money

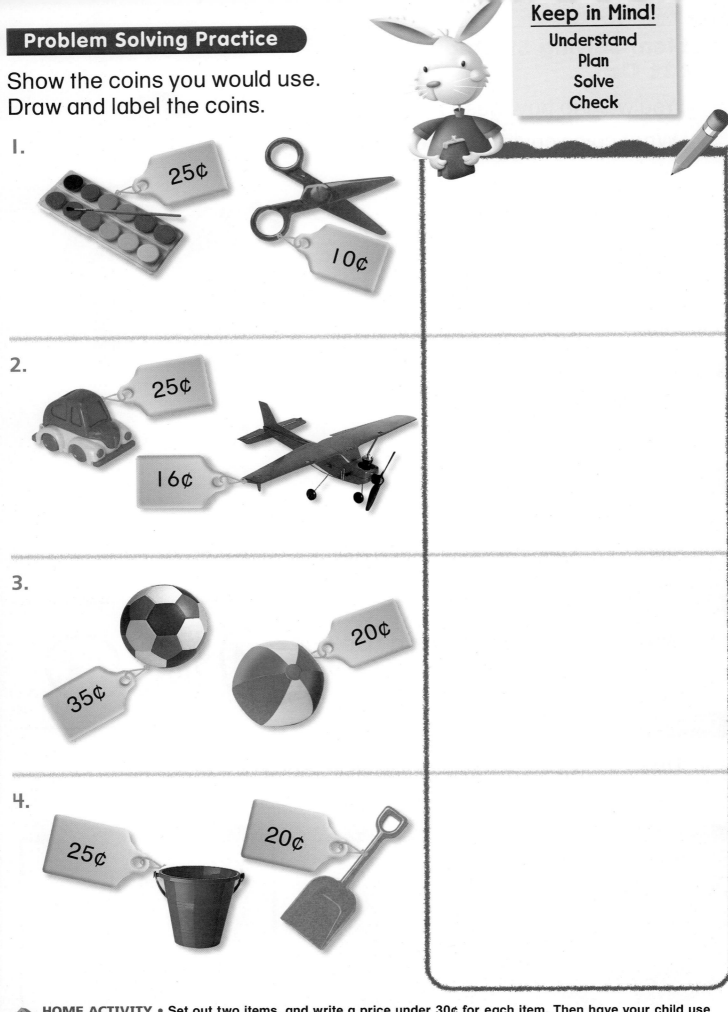

Name _____

Extra Practice

Use coins. Show the amount in two ways. Draw and label the coins. Circle the way that uses fewer coins.

1.

Count on from the quarter. Write the total amount.

2.

_____ ¢, _____ ¢, _____ ¢   ☐ ¢

Draw and label the coins. Write how many.

3. Show how many quarters equal 1 half dollar.

    ____ quarters = 1 half dollar

4. Show how many quarters equal 1 dollar.

    ____ quarters = 1 dollar

Problem Solving

Show the coins you would use. Draw and label the coins.

5.

Chapter 23 • Using Money   three hundred ninety-five **395**

Name _____

✓ Review/Test

Concepts and Skills

Use coins. Show the amount in two ways. Draw and label the coins. Circle the way that uses fewer coins.

1.

Count on from the quarter. Write the total amount.

2.

   _____¢, _____¢, _____¢, _____¢   ☐ ¢

Draw and label the coins. Write how many.

3. Show how many dimes equal 1 half dollar.

   _____ dimes = 1 half dollar

4. Show how many dimes equal 1 dollar.

   _____ dimes = 1 dollar

Problem Solving

Show the coins you would use. Draw and label the coins.

5.

396 three hundred ninety-six

Name _____

★ Standardized Test Prep
Chapters 1–23

Choose the answer for questions 1 – 4.

1. 5 + 3 = _____

   2   4   8   9
   ○   ○   ○   ○

2. How many [quarter] make [dollar bill] ?

   2   4   5   8
   ○   ○   ○   ○

3. Which is a way to make 25¢?

○   ○   ○   ○

4. Which amount can you trade to equal the amount shown?

○   ○   ○   ○

Show What You Know

5. Write a price on each toy that is less than 50¢. Draw coins to show each price. Explain which toy costs less. Circle it.

Chapter 23   three hundred ninety-seven **397**

Name _____

MATH GAME

Shopping Basket

Play with a partner.

1. Each player turns up 1 card.
2. Each player uses coins to show the total amount for the two cards.
3. The person who shows the correct amount with fewer coins keeps the two cards.
4. If both players use the same number of coins, each keeps one card.
5. Play until all the cards are used.
6. The player with more cards wins.

You will need

12 cards with prices

pile of

Stack the cards face down here.

Turn up 1 card here.

Turn up 1 card here.

CHAPTER 24 Telling Time

FUN FACTS

This clock has 2 hands like most clocks. It is special because it is made of foam material which does not break.

Theme: What Time Is It?

Name _____

✓ Check What You Know

More Time, Less Time

Circle the activity that takes more time.

1.

Circle the activity that takes less time.

2.

Use a Clock

Write the number that tells the hour.
Circle the two clocks that show the same time.

3.  4.  5.

_____ o'clock   _____ o'clock   _____ o'clock

Read a Clock

Explore

Write the missing numbers on the clock.

The time is 4 o'clock.

Vocabulary
- minute hand
- hour hand
- o'clock

Connect

Use a 🕐. Show each time.
Trace the hour hand. Write the time.

1.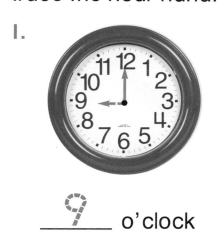

   __9__ o'clock

2.

   _____ o'clock

3.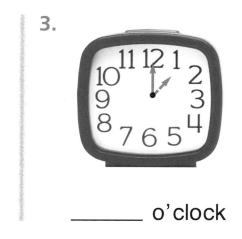

   _____ o'clock

Explain It • Daily Reasoning

How are the minute hand and the hour hand different?

Practice and Problem Solving

Use a 🕐. Show each time.
Trace the hour hand. Write the time.

1.

   __4__ o'clock

2.

   _____ o'clock

3.

   _____ o'clock

4.

   _____ o'clock

5.

   _____ o'clock

6.

   _____ o'clock

Problem Solving
Visual Thinking

Write the time.

7.

   _____ o'clock

8.

   _____ o'clock

 Write About It • Write about what you do at 2 o'clock on a school day.

⬡ **HOME ACTIVITY** • At times on the hour, have your child show you the minute hand and the hour hand on a clock and tell what time it is.

Name _____

Problem Solving Skill
Use Estimation

Vocabulary
minute

How long is a minute?
You can estimate it.

Close your eyes.

Estimate when 1 minute has passed. Raise your hand.

Was your estimate too long or too short?
Try again. Was your estimate closer this time?

About how long would it take?
Circle your estimate.
Then act it out to check.

1. snap your fingers

more than a minute
(less than a minute)

2. wave goodbye

more than a minute
less than a minute

3. write a story

more than a minute
less than a minute

Chapter 24 • Telling Time

four hundred three **403**

Problem Solving Practice

About how long would it take?
Circle your estimate.
Then act it out to check.

1. write 1 to 50

(more than a minute)

less than a minute

2. clap your hands

more than a minute

less than a minute

3. draw a picture

more than a minute

less than a minute

4. say your name

more than a minute

less than a minute

5. read a book

more than a minute

less than a minute

6. write your name

more than a minute

less than a minute

HOME ACTIVITY • Have your child name an activity that he or she thinks will take about one minute. Time the activity to see if it takes about one minute, more than one minute, or less than one minute.

Name _____

Time to the Hour

Vocabulary
hour

Explore

These clocks show time to the hour.

Both clocks show 9 o'clock.

Connect

Use a 🕐. Show each time. Write the time.

1. 12:00
2. :
3. :
4. :
5. :
6. :

Explain It • Daily Reasoning

How far does each clock hand move in one hour?

Practice and Problem Solving

Use a 🕐. Show each time. Write the time.

1.

2.

3.

4.

5.

6.

Problem Solving
Mental Math

Solve. Write the time.

7. Matt wakes up at 6 o'clock. Linda wakes up 1 hour later. What time does Linda wake up?

_____ o'clock

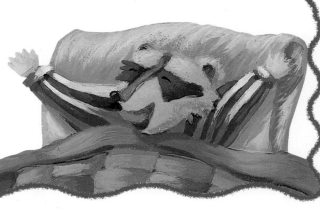

 Write About It • Look at Exercise 7. You wake up 3 hours later than Matt. What time is it? Tell how you know.

⬠ **HOME ACTIVITY** • Ask your child to say the times on the hour in order, beginning with 1 o'clock (1 o'clock, 2 o'clock, 3 o'clock, and so on).

Name _____

Tell Time to the Half Hour

Vocabulary
half hour

Explore

There are 60 minutes in an hour.

3:00 or
3 o'clock

There are 30 minutes in a half hour.

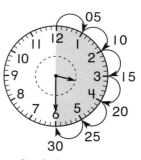

3:30 or
30 minutes after 3 o'clock

Connect

Use a 🕐 to show the time. Where are the hands? Write the numbers. Write the time.

1. The hour hand is between __2__ and __3__.

   The minute hand is at __6__.

   __2:30__

2.

   The hour hand is between _____ and _____.

   The minute hand is at _____.

   _____ : _____

3.

   The hour hand is between _____ and _____.

   The minute hand is at _____.

   _____ : _____

Explain It • Daily Reasoning

How far does each clock hand move in a half hour?

Chapter 24 • Telling Time
four hundred seven **407**

Practice and Problem Solving

Use a 🕐 to show the time.
Write the time.

1.

   12:30

2.

   ___ : ___

3.

   ___ : ___

4.

   ___ : ___

5.

   ___ : ___

6.

   ___ : ___

Problem Solving
Estimation

Estimate. Circle **half hour** or **hour**.
Then measure with a clock.

	Activity	Estimate.	Measure.
7.	 eat lunch	half hour hour	_____
8.	math class	half hour hour	_____

 Write About It • Write a list of things you can do in 30 minutes.

🏠 **HOME ACTIVITY** • At times on the half hour, have your child show you the minute hand and the hour hand on a clock and tell what time it is.

Name _____

Practice Time to the Hour and Half Hour

Learn

3:00

The hour hand points to a number. The minute hand points to 12.

3:30

The hour hand points half way between numbers. The minute hand points to 6.

Check

Draw the hour hand and the minute hand.

1.

   4:30

2.

   9:00

3.

   10:30

4.

   8:00

5.

   2:30

6.

   11:00

Explain It • Daily Reasoning

At 1:30, where is the hour hand? Explain.

Chapter 24 • Telling Time   four hundred nine **409**

Practice and Problem Solving

Draw the hour hand and the minute hand.

1.
   1:30

2.
   10:00

3.
   8:30

4.
   6:00

5.
   12:30

6.
   5:30

7.
   9:30

8.
   2:00

9.
   11:30

Problem Solving
Algebra

10. Continue the pattern.
    Write the times that are missing.

    1:00, 1:30, 2:00, ___:___, ___:___, 3:30, ___:___, ___:___

 Write About It • Look at Exercise 10.
What is the pattern?

HOME ACTIVITY • At times on the hour, have your child read the time on a clock and then tell what time it will be in half an hour.

Name _____

Extra Practice

Write the time.

1.

    _____ o'clock

2.

    _____ o'clock

3.

    _____ o'clock

4.

5.

6.

7.

    __ : __

8.

    __ : __

9.

    __ : __

Problem Solving

10. About how long would it take to write the alphabet? Circle your estimate.

    more than a minute

    less than a minute

Chapter 24 • Telling Time

four hundred eleven 411

Name _____

✓ Review/Test

Concepts and Skills
Write the time.

1.

_____ o'clock

2.

_____ o'clock

3.

_____ o'clock

4.

5.

6.

7.

___ : ___

8.

___ : ___

9.

___ : ___

Problem Solving

10. About how long would it take to write the numbers from 1 to 100? Circle your estimate.

    more than a minute

    less than a minute

Name _____

Standardized Test Prep
Chapters 1–24

Choose the answer for questions 1–3.

1. Which object has 1 face?

○   ○   ○   ○

2. Which takes less than 1 minute?

write your name   write a story   eat lunch   read a book

○   ○   ○   ○

3. Which clock shows the same time?

○   ○

○   ○

Show What You Know

4. Draw the hour hand and the minute hand on the clock.

Fill in the blanks to explain where the hands belong.

The hour hand is between

_____ and _____.

The minute hand is at _____.

Chapter 24   four hundred thirteen **413**

Name _____

MATH GAME

Clock Switch

Play with a partner.

1. One player uses ●. The other player uses ○.
2. Your partner picks any space.
3. You show that time on the other kind of clock.
4. If you are correct, put a counter there.
5. Play until all the spaces are covered.
6. The player with more counters wins.

You will need

25 ●   25 ○

CHAPTER 24 • MATH GAME

414   four hundred fourteen

CHAPTER 25
Time and Calendar

FUN FACTS

Helper dogs train for 4 months at their school. Then they train 1 more month with their new owner.

Theme: All in My Day

Name _____

✓ Check What You Know

Morning, Afternoon, and Evening

Identify the times of day.
Circle the time of day that is missing.

Use a Calendar: Identify Parts

DECEMBER						
Sunday	Monday	Tuesday	Wednesday	Thursday	Friday	Saturday
			1	2		4
5	6		8	9	10	11
12	13	14	15	16	17	
19	20	21	22	23	24	25
	27	28	29	30	31	

Fill in the missing numbers.
Circle the name of the month.
Color the first day of the month red.

Count the Mondays. _____   Count the Thursdays. _____

416   four hundred sixteen   Use this page to review important skills needed for this chapter.

Name _____

Use a Calendar

Vocabulary
month

Learn

These are the months of the year in order.

This month is May.

Check

Use the calendar to answer the questions.

1. How many days are in a week? _____ days

2. What day of the week is May 14? _____

3. How many days are in May? _____ days

4. What month comes before May? _____

5. What is the first month of the year? _____

Explain It • Daily Reasoning

How many months are left in the year after the month of May? Tell how you know.

Practice and Problem Solving

Fill in the calendar for next month.
Use the calendar to answer the questions.

Sunday	Monday	Tuesday	Wednesday	Thursday	Friday	Saturday

1. Color the Tuesdays 🖍 .

2. Color the Saturdays 🖍 .

3. What day of the week is the thirtieth? _____

4. On what day does the month end? _____

5. What is the date of the first Thursday? _____

6. How many days are in the month? _____ days

Problem Solving
Application

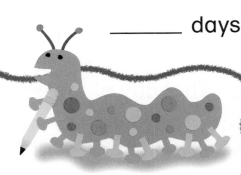

Look at the calendar above.
Suppose today is the twentieth.

7. What day of the week was yesterday? Circle it in 🖍 .

8. What day of the week is tomorrow? Circle it in 🖍 .

 Write About It • Write about your favorite day of the week. Tell why it is your favorite.

⬟ **HOME ACTIVITY** • Have your child draw pictures of special events in his or her life and label each with the month of the year in which it happens.

Name _____

Daily Events

Learn

Vocabulary
morning
afternoon
evening

I do this in the morning.

I do this in the afternoon.

I do this in the evening.

Check

Draw pictures of things you do in the morning, in the afternoon, and in the evening.

Morning

Afternoon

Evening

Explain It • Daily Reasoning

How are the things you do in the morning different from the things you do in the evening?

Chapter 25 • Time and Calendar

four hundred nineteen **419**

Is It Time?

written by Lucy Floyd
illustrated by Liz Conrad

🏠 This book will help me review telling time.

This book belongs to _____.

It can't be time so soon!
Tick-tock. Tick-tock. Tick-tock.
Do I need to get up now?

Yes! It's _____ o'clock!

_____ : _____

A half hour later . . .

I like to walk to school.
I walk just one short block.
Is it time for school to start?

Yes! It's ____ o'clock!

A half hour later . . .

We do our math in school until we hear a knock. Is it time to go to art?

Yes! It's ____ o'clock!

☐ : ☐

A half hour later . . .

When school is out, we play.
I show a friend my rock.
Is it time for me to go?

Yes! It's ____ o'clock!

__ : __

A half hour later . . .

Dad cooks some beans for us.
He stirs them in a wok.
Is it time for supper now?

Yes! It's _____ o'clock!

A half hour later . . .

I sit with Mom at night.
We like to sing and rock.
Is it time to read a book?

Yes! It's _____ o'clock!

A half hour later . . .

Now it's time to
go to sleep . . .

Tick-tock. Tick-tock.

PROBLEM SOLVING ON LOCATION

At the Jazz Festival

New Orleans, Louisiana, has a jazz festival each year. People come to enjoy the music.

The musical groups ♥, ◆, and ■ are playing at the festival.

Use the calendar to answer the questions.

APRIL

Sunday	Monday	Tuesday	Wednesday	Thursday	Friday	Saturday
					1	2
3	4	5	6	7	8 ♥	9
10	11	12 ◆	13	14	15 ■	16
17	18	19	20	21	22	23
24	25	26	27	28	29	30

1 On what day of the week is ◆ playing? _____

2 Color the day of the week that has two groups playing.

3 Which group is playing on April 8?

Unit 5 • Chapters 21–25   four hundred thirty-one **431**

CHALLENGE

Writing Fractions

What fraction is shaded?

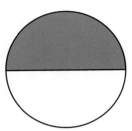
1 part is shaded out of
2 equal parts.

 1 part
 3 equal parts

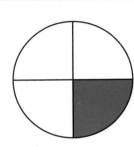 1 part
4 equal parts

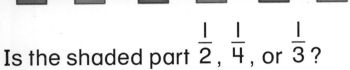

Is the shaded part $\frac{1}{2}$, $\frac{1}{4}$, or $\frac{1}{3}$? Write the fraction.

1.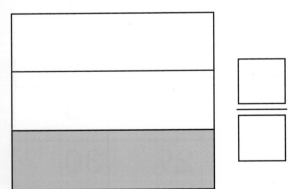

2.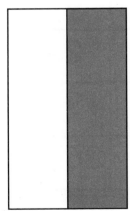

3.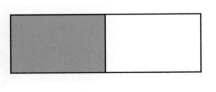

4.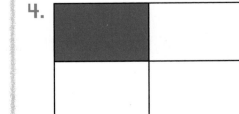

432 four hundred thirty-two

✓ Study Guide and Review

Vocabulary

Circle the **penny** 🖍. Circle the **dime** 🖍.
Circle the **quarter** 🖍. Circle the **nickel** 🖍.

1.

Skills and Concepts

Count the coins. Write the amount.

2.

   _____ ¢, _____ ¢, _____ ¢, _____ ¢, _____ ¢   ☐ ¢

3.

   _____ ¢, _____ ¢, _____ ¢, _____ ¢, _____ ¢   ☐ ¢

Use coins. Show the amount in two ways. Draw and label the coins. Circle the way that uses fewer coins.

4.

Draw and label the coins. Write how many.

5. Show how many dimes equal 1 dollar.

   _____ dimes = 1 dollar

Unit 5 • Study Guide and Review   four hundred thirty-three **433**

Write the time.

6.

7.

8.

Color one part. Circle the fraction for the shaded part.

9.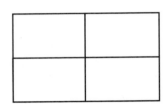

$\frac{1}{2}$ $\frac{1}{3}$ $\frac{1}{4}$

10.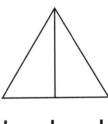

$\frac{1}{2}$ $\frac{1}{3}$ $\frac{1}{4}$

11.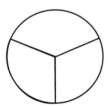

$\frac{1}{2}$ $\frac{1}{3}$ $\frac{1}{4}$

Problem Solving

Circle the best estimate for the activity.

12. Take a walk in the park.

about one minute

about one hour

about one week

434 four hundred thirty-four

Name _____

✓ Performance Assessment

Party Favors

Ed had these hats and blowers to give to friends at his party.

Choose one of the objects in the picture.

Use a fraction to tell what part of a group the object is.

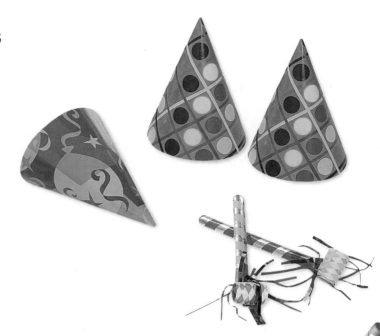

Show your work.

Draw a picture to show the group.
Color the object you chose.

What part of the group does your fraction show?

Unit 5 • Performance Assessment

four hundred thirty-five **435**

Name _____

TECHNOLOGY

The Learning Site • Willy the Watchdog

1. Go to **www.harcourtschool.com**.
2. Click on .
3. Play with a friend.
4. Set the clock to match each time.
5. Play again.

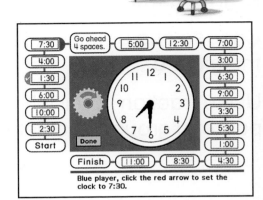

Practice and Problem Solving

Write the time.

1.

   3:00

2.

   ___ : ___

3.

   ___ : ___

Draw the hour hand and the minute hand.

4.

   11:00

5.

   2:30

6.

   6:30

7. Draw the hands on the clock to show 30 minutes later than 5:00.

436 four hundred thirty-six

LOOKING BACK
SCHOOL HOME CONNECTION

Dear Family,
  In Unit 5 we learned about money, time, and fractions. Here is a game for us to play together. This game will give me a chance to share what I have learned.

  Love,
  _____

Directions
1. Put your game piece at START.
2. Your partner hides 3 pennies in one hand and 2 pennies in the other.
3. Tap on one of your partner's hands. Your partner opens that hand.
4. Count the pennies. Move forward that many spaces.
5. Tell about the picture you land on.
   - Show the value with coins.
   - Tell what fraction is shown.
   - Tell what time the clock shows.
6. Take turns.
7. The first person to get to END wins.

Materials
- pile of pennies, nickels, dimes, and quarters
- 2 game pieces or 2 rocks

Tell About It

Unit 6 • Unit Game

four hundred thirty-seven A **437A**

SCHOOL HOME CONNECTION
LOOKING FORWARD

Dear Family,

During the next few weeks, we will learn about measurement. We will also learn about adding and subtracting 2-digit numbers. Here is important math vocabulary and a list of books to share.

Love,

_____

Vocabulary
- inch
- foot
- pound
- cup

Vocabulary Power

An **inch** is a customary unit for measuring short lengths.

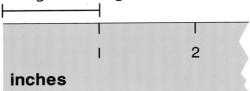

inches

A **foot** is a customary unit for measuring longer lengths.

This bag of sugar weighs 1 **pound**.

A **cup** is a customary unit for measuring how much an object holds.

BOOKS TO SHARE

To read about measurement and about 2-digit numbers with your child, look for these books in your library.

Koala Lou, by Mem Fox, Harcourt, 1994.

Stone Soup, by Ann McGovern, Scholastic, 1986.

Lulu's Lemonade, by Barbara deRubertis, Kane Press, 2000.

One Hundred Hungry Ants, by Elinor J. Pinczes, Houghton Mifflin, 1999.

 Visit *The Learning Site* for additional ideas and activities. www.harcourtschool.com

CHAPTER 26 Length

FUN FACTS

You can make a 100 day paper chain by cutting 10 strips of paper all 1 inch by 9 inches in 10 different colors.

Theme: Arts and Crafts

✓ Check What You Know

Compare Length

Circle the longer object.
Draw a line under the shorter object.

1.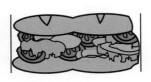

   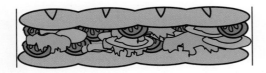

2.

3.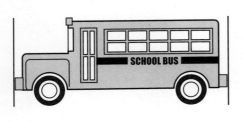

   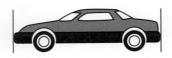

4.

   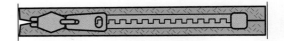

Order Length

Circle the objects that are in order from shortest to longest, starting at the top.

5.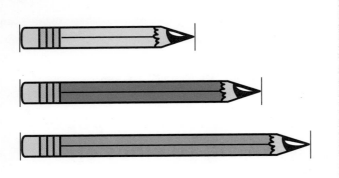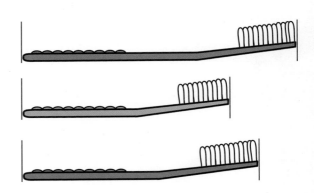

438 four hundred thirty-eight   Use this page to review important skills needed for this chapter.

Name _____

Compare Lengths

Vocabulary
shortest
longest

Hands On Explore

These paper strips are in order from shortest to longest.

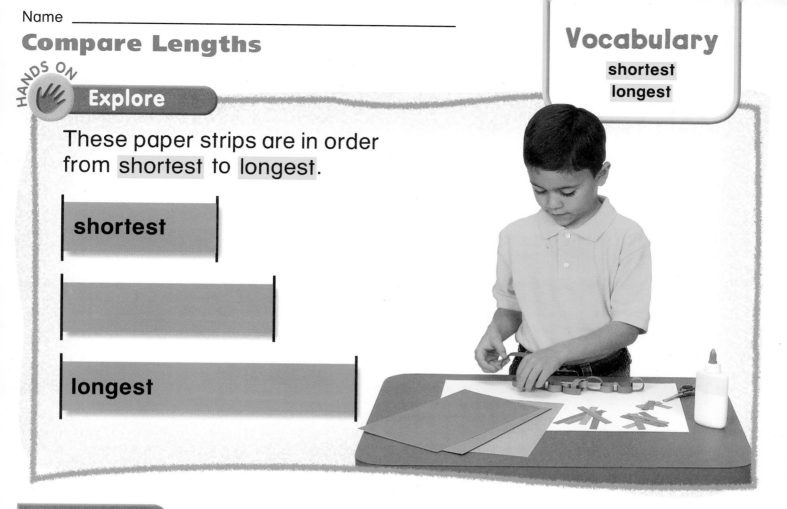

Connect

Put three paper strips in order from shortest to longest. Draw them.

1. shortest

2.

3. longest

Explain It • Daily Reasoning

In what other way could you put the paper strips in order?

Chapter 26 • Length

four hundred thirty-nine **439**

Practice and Problem Solving

Use real objects. Cut yarn to show each length. Use different colors. Then compare the pieces of yarn. Tell which object is the longest and which is the shortest.

Circle the object to answer the question.

1. Which is longer?

2. Which is shorter?

3. Which is longer?

4. Which is the longest?

Problem Solving
Visual Thinking

5. Circle the string that is longer. Use real string to check.

 Write About It • Look at Exercise 5. Explain your answer.

HOME ACTIVITY • Give your child three small objects of different lengths. Ask him or her to put them in order from shortest to longest.

Name _____

Use Nonstandard Units

Explore

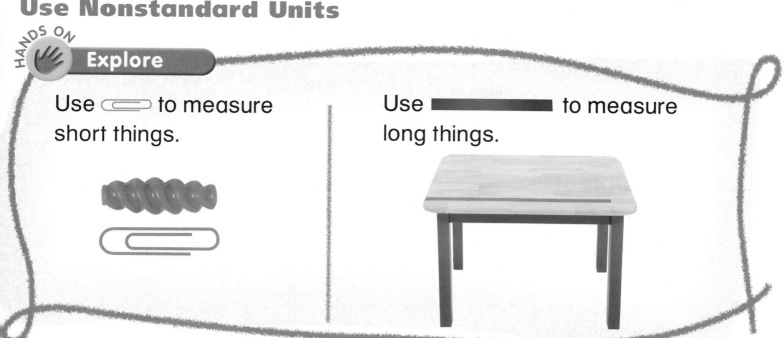

Use 🖇 to measure short things.

Use ▬▬▬ to measure long things.

Connect

Use real objects. Circle the unit you would use to measure. Then measure.

	Object	Unit	Measurement
1.	desk	🖇 / ▬▬▬	about _____
2.	scissors	🖇 / ▬▬▬	about _____
3.	bookshelf	🖇 / ▬▬▬	about _____
4.	pencil	🖇 / ▬▬▬	about _____

Explain It • Daily Reasoning

How do you decide which unit to choose?

Chapter 26 • Length

four hundred forty-one **441**

Name _____

Extra Practice

Circle the string that is the shortest.

1.

Use the real object and . Estimate. Then measure.

2.

Estimate about _____

Measurement about _____

Use the real object and an inch ruler. Estimate. Then measure.

3.

Estimate about _____ inches

Measurement about _____ inches

Use the real object and a centimeter ruler. Estimate. Then measure.

4.

Estimate   about _____ centimeters

Measurement   about _____ centimeters

Problem Solving

About how many beads long is the string?
Circle the answer that makes sense.

5.

about 2   about 6   about 10

Chapter 26 • Length   four hundred fifty-one **451**

Name _____

✅ Review/Test

Concepts and Skills

Circle the string that is the longest.

1.

Use the real object and ⌒. Estimate. Then measure.

Use the real object and an inch ruler. Estimate. Then measure.

2.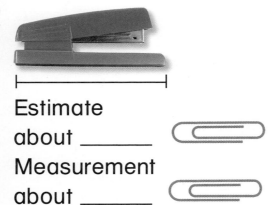

Estimate
about _____ ⌒

Measurement
about _____ ⌒

3.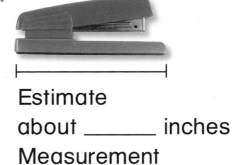

Estimate
about _____ inches

Measurement
about _____ inches

Use the real object and a centimeter ruler. Estimate. Then measure.

4.

Estimate   about _____ centimeters

Measurement   about _____ centimeters

Problem Solving

About how many beads long is the string?
Circle the answer that makes sense.

5.

about 3   about 5   about 7

452 four hundred fifty-two

Name _____

★Standardized Test Prep
Chapters 1–26

Choose the answer for questions 1–5.

1. Which crayon is the shortest?

2. Use a 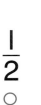 to measure.
   About how long is the paint brush?

    2   3   4   8
    ○   ○   ○   ○

3. What fraction does the green part show?

   $\frac{1}{2}$ $\frac{1}{3}$ $\frac{1}{4}$ $\frac{1}{5}$
   ○   ○   ○   ○

4. Which unit would you use to measure a door?

   ○ inch

   ○ centimeter

   ○ foot

5. Which is the best estimate?
   About how many beads long is the string?

    4   5   6   7
    ○   ○   ○   ○

Show What You Know

6. Draw a ruler that is 3 inches long. Mark each inch. Use your ruler to measure your thumb. Draw a line to show how long it is.

   Is your thumb longer or shorter than your ruler? Circle to explain.

MATH GAME

Ruler Race

Play with a partner.

1. Put your 🔺 at START.
2. Spin the 🎯.
3. Use a ruler to find an object about that many inches long.
4. Move your 🔺 that many spaces.
5. The first player to get to END wins.

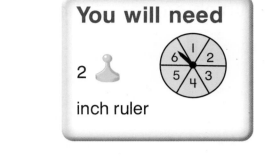

You will need

2 🔺

inch ruler

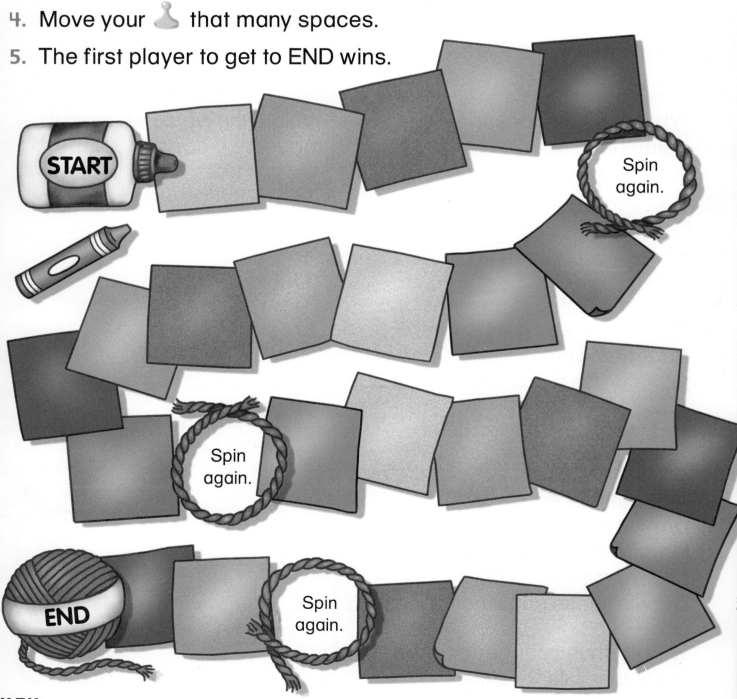

CHAPTER 27 Weight

FUN FACTS

SOCIAL STUDIES

Baby chicks weigh about 1 pound, which is about the same as 2 large apples.

Theme: At the Farm

Name _____

✓ Check What You Know

Compare Weight

Circle the object that is heavier.
Mark an X on the object that is lighter.

1.

2.

3.

4.

5.

6.

456 four hundred fifty-six   Use this page to review important skills needed for this chapter.

Name _____

Use a Balance

Vocabulary
balance

HANDS ON Explore

"It takes 6 bears to **balance** the crayons."

"It takes a lot of paper clips to balance the crayons."

Connect

Use a  and real objects.
Circle the unit you would use to measure.
Then measure.

	Object	Unit	Measurement
1.	scissors	paper clip / bear	about _____
2.	dime	paper clip / bear	about _____
3.	pencil sharpener	paper clip / bear	about _____
4.	glue bottle	paper clip / bear	about _____

Explain It • Daily Reasoning

How did you decide which unit to use?

Chapter 27 • Weight

four hundred fifty-seven **457**

Practice and Problem Solving

About how many ⌒ does it take to balance?
Use real objects, a ⚖, and ⌒.
Estimate. Then measure.

Object	Estimate	Measurement
1. ✏️	about _____ ⌒	about _____ ⌒
2. 📏	about _____ ⌒	about _____ ⌒
3. ▭	about _____ ⌒	about _____ ⌒
4. 🧽	about _____ ⌒	about _____ ⌒

5. Circle the heaviest object in blue.
6. Circle the lightest object in red.

Problem Solving

Logical Reasoning

7. 2 boxes of ⌒ balance a 🥮.

   6 boxes of ⌒ balance a 📕.

   How many 🥮 will balance the 📕? _____ 🥮

Write About It • Look at Exercise 5.
Draw a picture to show your work.

 HOME ACTIVITY • Give your child some cans or boxes. Ask him or her to put them in order from heaviest to lightest.

Name _____

Pounds

Vocabulary
pound

Explore (Hands On)

This bag of sugar weighs about one pound.

This bag of flour weighs about 5 pounds.

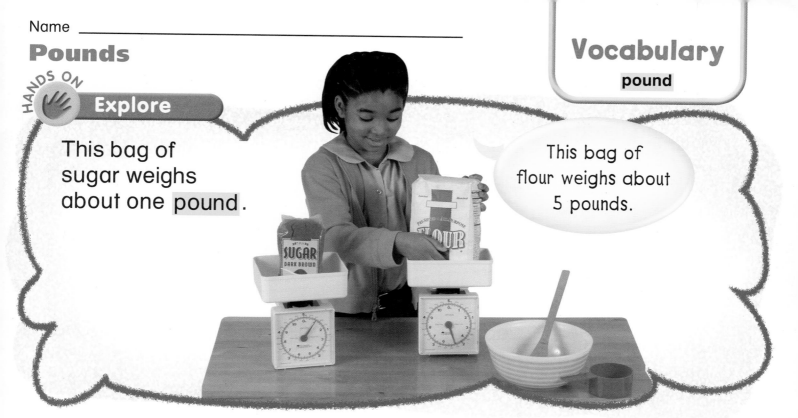

Connect

Look at each object.
Circle the better estimate.

1.

   (about 1 pound)

   about 10 pounds

2.

   about 1 pound

   about 10 pounds

3.

   about 1 pound

   about 10 pounds

4.

   about 1 pound

   about 10 pounds

Explain It • Daily Reasoning

Explain how you estimated what the objects would weigh.

Chapter 27 • Weight

Practice and Problem Solving

Find three items to weigh. Draw them. Estimate how much each object weighs. Then measure.

	Object	Estimate	Measurement
1.		about _____ pounds	about _____ pounds
2.		about _____ pounds	about _____ pounds
3.		about _____ pounds	about _____ pounds

4. Circle the heaviest item in blue.
5. Circle the lightest item in red.

Problem Solving
Visual Thinking

Does each object weigh more than or less than 1 pound? Circle the better estimate.

6.

| more than 1 pound | more than 1 pound | more than 1 pound |
| less than 1 pound | less than 1 pound | less than 1 pound |

Write About It • Draw three things that weigh less than a pound. Tell how you know.

HOME ACTIVITY • Ask your child to read you the weights, in pounds, of some grocery items.

Name _____

Kilograms

Hands On Explore

Vocabulary
kilogram
gram

This large book is about 1 kilogram.

This paper clip is about 1 gram.

Connect

Circle the unit you would use to measure the real object.

1. (grams) kilograms	2. grams kilograms	3. grams kilograms
4. grams kilograms	5. grams kilograms	6. grams kilograms
7. grams kilograms	8. grams kilograms	9. grams kilograms

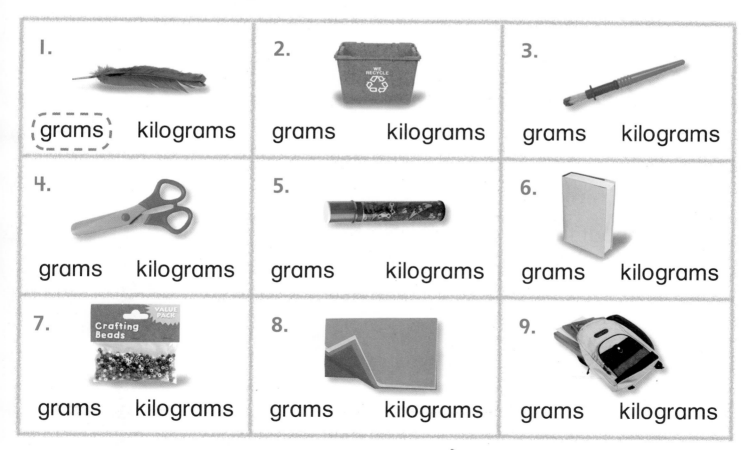

Explain It • Daily Reasoning

What kind of objects did you choose to measure in grams? Why?

Chapter 27 • Weight

four hundred sixty-one **461**

Practice and Problem Solving

Estimate how much the real object will measure. Use grams or kilograms. Then measure.

Object	Estimate	Measurement
1.	about _____ kilograms	about _____ kilograms
2.	about _____ grams	about _____ grams
3.	about _____ kilograms	about _____ kilograms
4.	about _____ grams	about _____ grams

Problem Solving
Visual Thinking

Think about the real objects. Which would you measure in grams? Circle in crayon.

5.

Write About It • Draw three things that weigh more than 1 kilogram. Tell how you know.

HOME ACTIVITY • Ask your child to read you the weights, in grams, of some grocery items.

Name _____

Problem Solving Strategy
Predict and Test

How many grams is this marker?

UNDERSTAND

What do you want to find out?
How many grams is the marker?

PLAN

How will you solve the problem?
I will predict how many grams it is.
To test, I will use the balance to measure.

SOLVE

Predict.
Then use large 🖇 as grams to balance.

THINK:
A large paper clip is about 1 gram.

Predict _____ grams   Test __9__ grams

CHECK

Was your prediction close?
Explain.

How many grams is the object?
Use the real object and 🖇.

Predict. Then test.

1.

Predict _____ grams   Test _____ grams

Chapter 27 • Weight

four hundred sixty-three **463**

Problem Solving Practice

How many grams is the object?
Use the real object and ⌒.
Predict. Then test.

Keep in Mind!
Understand
Plan
Solve
Check

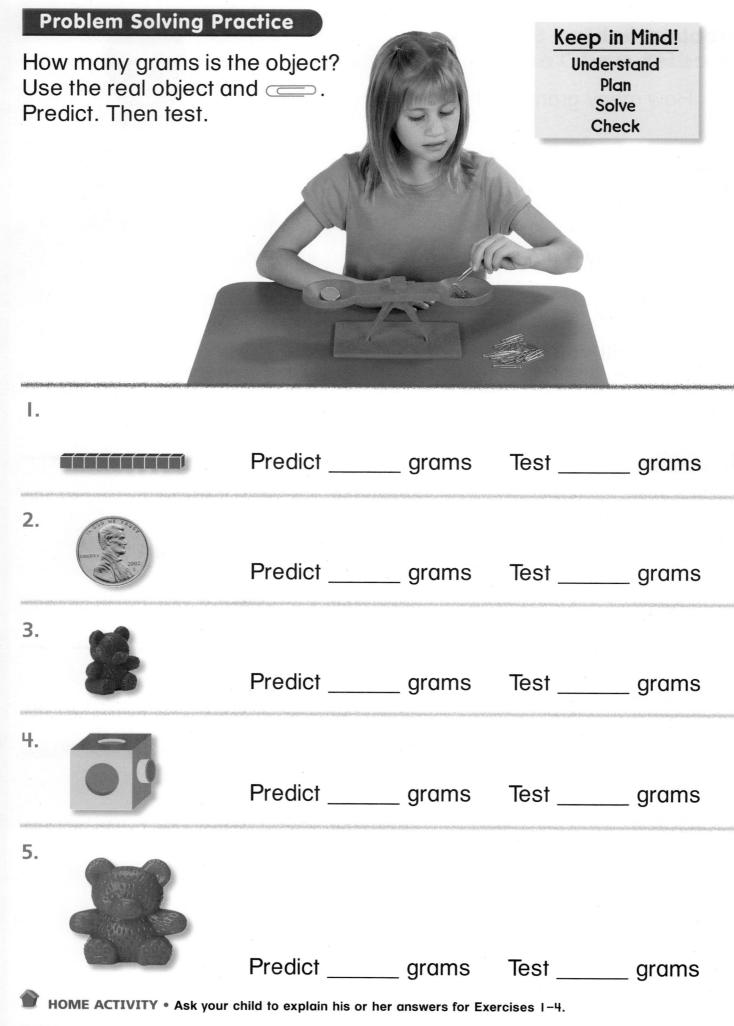

1. Predict _____ grams   Test _____ grams

2. Predict _____ grams   Test _____ grams

3. Predict _____ grams   Test _____ grams

4. Predict _____ grams   Test _____ grams

5. Predict _____ grams   Test _____ grams

🏠 **HOME ACTIVITY** • Ask your child to explain his or her answers for Exercises 1–4.

Name _____

Extra Practice

1. Choose the unit you would use to measure.
   Circle 🖇 or 🧸.
   Use a ⚖ to measure the real object.

Object	Unit	Measurement
(sharpener)	🖇 🧸	about _____

2. Find an object to weigh. Draw it.
   Estimate how much the object weighs.
   Then measure.

Object	Estimate	Measurement
	about _____ pounds	about _____ pounds

3. Estimate how much the real object will measure.
   Then measure.

Object	Estimate	Measurement
(bottle)	about _____ kilograms	about _____ kilograms

Problem Solving

4. How many grams is the object?
   Use the real object and 🖇.
   Predict. Then test.

Predict _____ grams   Test _____ grams

Chapter 27 • Weight   four hundred sixty-five **465**

Name _____

 Review/Test

Concepts and Skills

1. Choose the unit you would use to measure.
   Circle 📎 or 🧸.
   Use a ⚖ to measure the real object.

Object	Unit	Measurement
👟	📎 🧸	about _____

2. Find an object to weigh. Draw it.
   Estimate how much the object weighs.
   Then measure.

Object	Estimate	Measurement
	about _____ pounds	about _____ pounds

3. Estimate how much the object measures.
   Then measure.

Object	Estimate	Measurement
📕	about _____ kilograms	about _____ kilograms

Problem Solving

4. How many grams is the object?
   Use the real object and 📎.
   Predict. Then test.

Predict _____ grams   Test _____ grams

466   four hundred sixty-six

Name _____

★Standardized Test Prep
Chapters 1–27

Choose the answer for questions 1–4.

1. 20
   −10

   9   10   11   29
   ○   ○   ○   ○

2. Which object weighs about 1 kilogram?

○   ○   ○   ○

3. Which object weighs about 1 pound?

○   ○   ○   ○

4. Which is the heaviest?

1 gram   5 grams   5 grams   10 grams
○   ○   ○   ○

Show What You Know

5. How many grams is the object? Predict. Then test.

Predict _____ grams

Test _____ grams

Was your prediction too high, too low, or the same? Explain.

_____

Chapter 27   four hundred sixty-seven   **467**

MATH GAME

Gram Grab

Play with a partner.

1. Put your 🎯 at START.
2. Toss the 🎲.
3. Move your 🎯 that many spaces.
4. Measure the object shown on that space in grams.
5. Take that many ●.
6. When both players get to END, count ●.
7. The player with more ● wins.

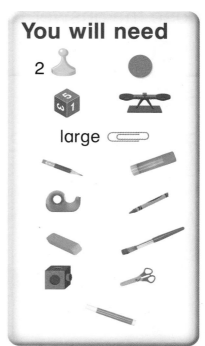

You will need

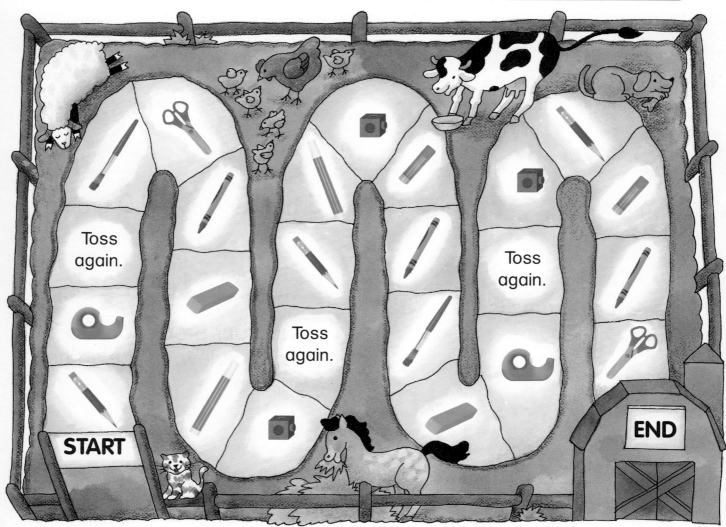

CHAPTER 28 Capacity

FUN FACTS

Each starfish needs 30 liters of saltwater to live in.

Theme: Fill It Up

Name _____

✓ Check What You Know

Compare Capacity

Circle the container that holds more.
Mark an X on the container that holds less.

1.

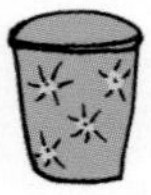

2.

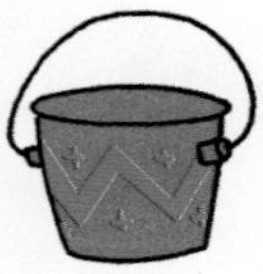

3.

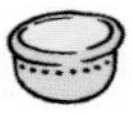

4.

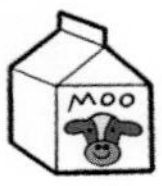

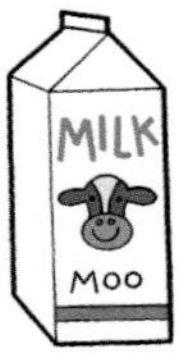

5.

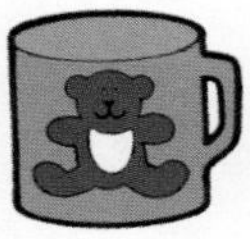

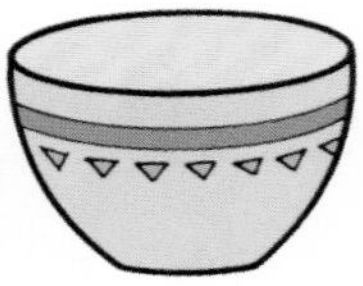

6.

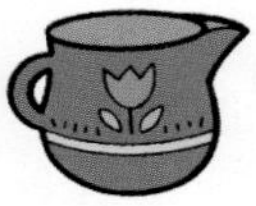

Name _____

Nonstandard Units

Explore

It took 3 of these scoops to fill the container.

It took 6 of these cups to fill the container.

Connect

Choose the unit you would use to measure.
Circle 🥄 or ☕. Measure.

	Container	Unit	Measurement
1.	Yogurt	🥄 ☕	about _____
2.	Milk carton	🥄 ☕	about _____
3.	Cottage Cheese	🥄 ☕	about _____
4.	Rice box	🥄 ☕	about _____

Explain It • Daily Reasoning

How did you decide which units to use?

Chapter 28 • Capacity

four hundred seventy-one 471

Practice and Problem Solving

Use the real container and a 🥄.
Estimate. Then measure.

	Container	Estimate	Measurement
1.		about _____ 🥄	about _____ 🥄
2.		about _____ 🥄	about _____ 🥄
3.		about _____ 🥄	about _____ 🥄
4.		about _____ 🥄	about _____ 🥄

Problem Solving

Estimation

Circle your answer.

5. Which container do you think will hold the most 🥄 of rice?

 Write About It • Look at Exercise 5. Explain how you chose your answer.

🏠 **HOME ACTIVITY** • Show your child three containers. Ask him or her to estimate which will hold the most water. Together, use a small cup or scoop to find out.

Name _____

Cups, Pints, and Quarts

Explore

You can use a cup to measure how much a container holds.

"This pint container will hold 2 cups."

"This quart container will hold 4 cups."

Vocabulary
cup
pint
quart

Connect

Use a 🥣 and containers.
Estimate. Then measure.

	Container	Estimate	Measurement
1.	(pot)	about _____ cups	about _____ cups
2.	(half & half carton)	about _____ cups	about _____ cups
3.	(measuring cup)	about _____ cups	about _____ cups
4.	(milk carton)	about _____ cups	about _____ cups

Explain It • Daily Reasoning

Which container holds about 2 cups?
Which containers hold about 4 cups?

Chapter 28 • Capacity   four hundred seventy-three **473**

Practice and Problem Solving

Estimate. Then measure.
Trace to name the size of the container.

	Container	Estimate	Measurement	Size
1.		about _____ cups	about _____ cups	pint
2.		about _____ cups	about _____ cups	pint
3.		about _____ cups	about _____ cups	quart

Problem Solving
Logical Reasoning

4. Circle what you would use to measure a quart of milk.

5. Circle what you would use to measure a cup of juice.

 Write About It • Look at Exercises 4 and 5. Explain how you decided which unit to use to measure.

HOME ACTIVITY • Show your child three containers, and ask him or her to estimate which will hold more than a pint. Together, use a cup measure to find out.

Name _____

Liters

Vocabulary
liter

Explore

This **liter** bottle will hold a little more than 4 cups.

This liter bottle will hold a little more than 1 quart.

Connect

Estimate whether the container holds less than or more than a liter. Then use a liter bottle to measure.

Container	Estimate	Measurement
1. (juice bottle)	less than a liter / more than a liter	less than a liter / more than a liter
2. (gallon jug)	less than a liter / more than a liter	less than a liter / more than a liter
3. (bowl)	less than a liter / more than a liter	less than a liter / more than a liter
4. (milk carton)	less than a liter / more than a liter	less than a liter / more than a liter

Explain It • Daily Reasoning

Would you use a liter bottle to fill a fish tank with water? Why or why not?

Chapter 28 • Capacity

four hundred seventy-five **475**

Practice and Problem Solving

Does the container hold less than or more than 2 liters? Estimate. Then measure.

Container	Estimate	Measurement
1.	less than 2 liters / more than 2 liters	less than 2 liters / more than 2 liters
2.	less than 2 liters / more than 2 liters	less than 2 liters / more than 2 liters
3.	less than 2 liters / more than 2 liters	less than 2 liters / more than 2 liters

Problem Solving
Logical Reasoning

4. A liter bottle holds a little more than 4 cups. About how many cups will there be in 2 liter bottles?

   about _____ cups

Write About It • Look at Exercise 4. Explain how you got your answer.

HOME ACTIVITY • Help your child find a 1-liter container and a 1-quart container. Ask him or her to tell you if a liter is less than or more than a quart. Then use water to check.

Name _____

Temperature

Vocabulary
temperature

Learn

A thermometer measures temperature.

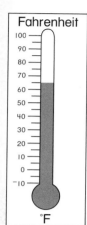

The temperature is 65 degrees.

It is __85__ °F.

It is __50__ °F.

Check

Read the thermometer.
Write the temperature.

1. 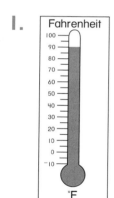   _____ °F

2. 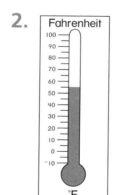   _____ °F

3.    _____ °F

4.    _____ °F

Explain It • Daily Reasoning

What happens to a thermometer when the temperature gets warmer?

Chapter 28 • Capacity

Practice and Problem Solving

Read the temperature.
Color the thermometer to show the temperature.

1. 60°F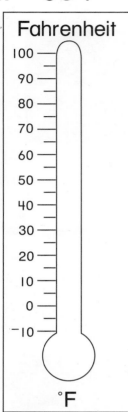
2. 20°F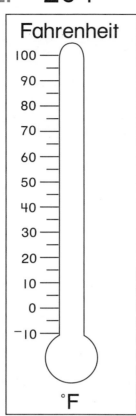
3. 75°F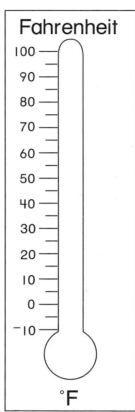
4. 95°F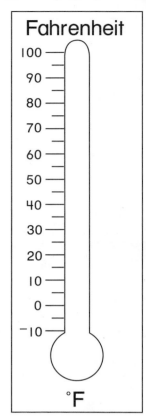

Problem Solving
Visual Thinking

5. The thermometer shows the temperature is 80 degrees. The sun goes down, and the temperature gets cooler. Circle the thermometer that shows what the temperature might be now.

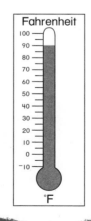 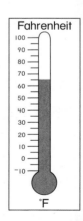

 Write About It • Tell what you would wear if the temperature was 85 degrees and what you would wear if it was 40 degrees.

HOME ACTIVITY • With your child, look at an outdoor thermometer or a weather report. Talk together about what the temperature is today.

Name _____

Problem Solving Skill
Choose the Measuring Tool

I use different tools to measure in different ways.

Find five objects to measure in different ways.
Choose the correct tool to measure in each way.
Draw and write to complete the chart.

	What to Find Out	Object	Tool	Measurement
1.	How tall is it?			
2.	How wide is it?			
3.	How much does it hold?			
4.	How much does it weigh?			
5.	How hot or cold is it?			

Chapter 28 • Capacity

Problem Solving Practice

Circle the correct tool to measure.

1. How tall is the plant?

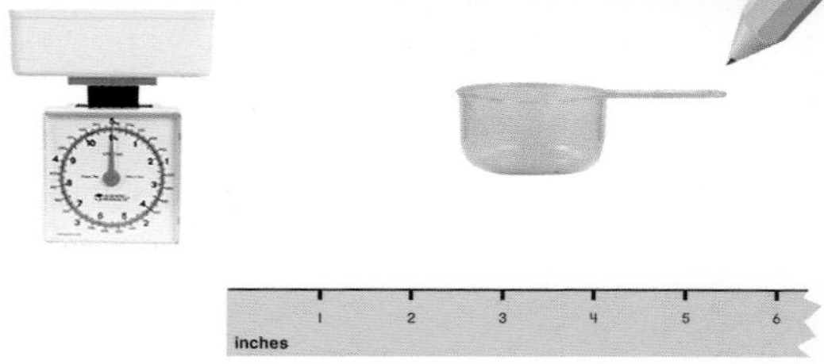

2. Which book is heavier?

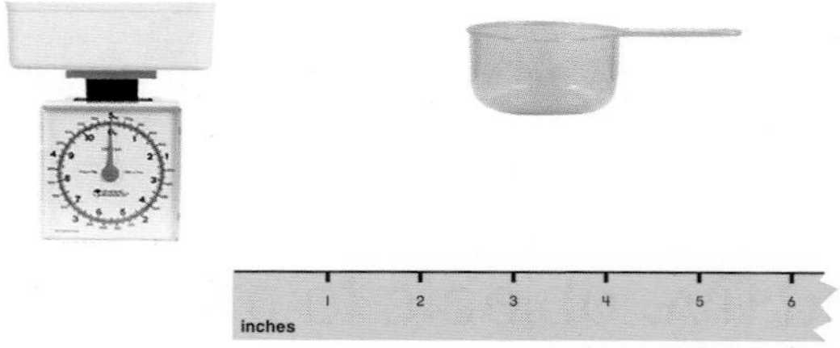

3. How much will the jar hold?

4. How wide is the chair?

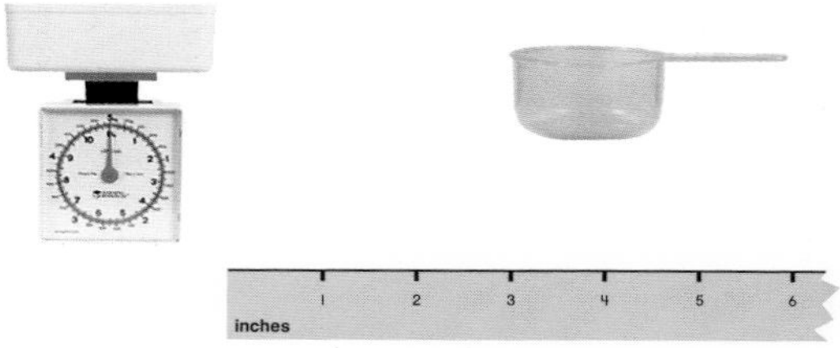

5. Which container holds more?

6. How cold is the water?

HOME ACTIVITY • Give your child an object, such as a book, and ask him or her to tell about all the different ways to measure it.

480 four hundred eighty

Name _____

Extra Practice

1. Use the real container and a 🥄.
Estimate. Then measure.

Container	Estimate	Measurement
[cottage cheese tub]	about _____ 🥄	about _____ 🥄

2. Estimate how many cups it will take to fill the container. Then measure.

Container	Estimate	Measurement
[cup]	about _____ cups	about _____ cups

3. Estimate whether the container holds less than or more than a liter. Then use a liter bottle to measure.

Container	Estimate	Measurement
[thermos]	less than a liter more than a liter	less than a liter more than a liter

Read the temperature. Color the thermometer to show the temperature.

4. 80°F

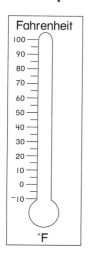

Problem Solving

5. Circle the correct tool to measure how hot the water is.

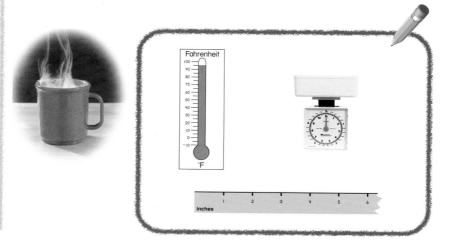

Chapter 28 • Capacity

four hundred eighty-one **481**

Name _____

✓ Review/Test

Concepts and Skills

1. Use the real container and a 🥄.
   Estimate. Then measure.

Container	Estimate	Measurement
(oats)	about _____ 🥄	about _____ 🥄

2. Estimate how many cups it will take to fill the container. Then measure.

Container	Estimate	Measurement
(half & half)	about _____ cups	about _____ cups

3. Estimate whether the container holds less than or more than a liter. Then use a liter bottle to measure.

Container	Estimate	Measurement
(bottle)	less than a liter more than a liter	less than a liter more than a liter

Read the temperature. Color the thermometer to show the temperature.

4. 45°F

Problem Solving

5. Circle the correct tool to measure how tall the notebook is.

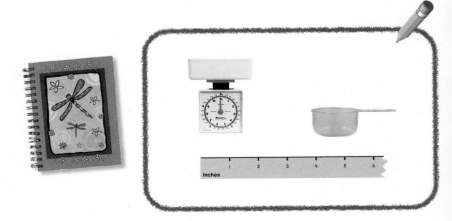

482 four hundred eighty-two

Name _____

★Standardized Test Prep
Chapters 1-28

Choose the answer for questions 1–5.

1. 15 − 8 = _____

   6 ○   7 ○   8 ○   9 ○

2. Which one would you use a small 🥄 to measure?

   ○   ○   ○   ○

3. Which container holds about 1 cup?

   ○   ○   ○   ○

4. Which container holds more than 1 quart?

   ○   ○   ○   ○

5. What is the temperature?

   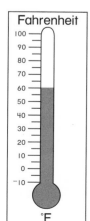

   20°F ○   32°F ○

   60°F ○   90°F ○

Show What You Know

6. How much will the glass hold? Circle the correct tool to explain how you would measure.

Chapter 28   four hundred eighty-three **483**

Name _____

MATH GAME

How Many Cups?

Play with a partner.

1. Put your 🔸 at START.
2. Spin the ⊘.
3. Move your 🔸 to the next space that matches that color.
4. Measure to find how many cups that container holds.
5. Take 1 ● for each cup it holds.
6. Take turns until both players get to END.
7. The player with more counters wins.

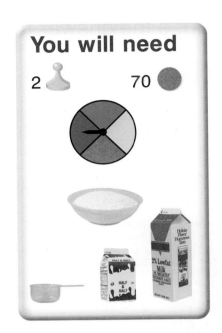

You will need

2 🔸   70 ●

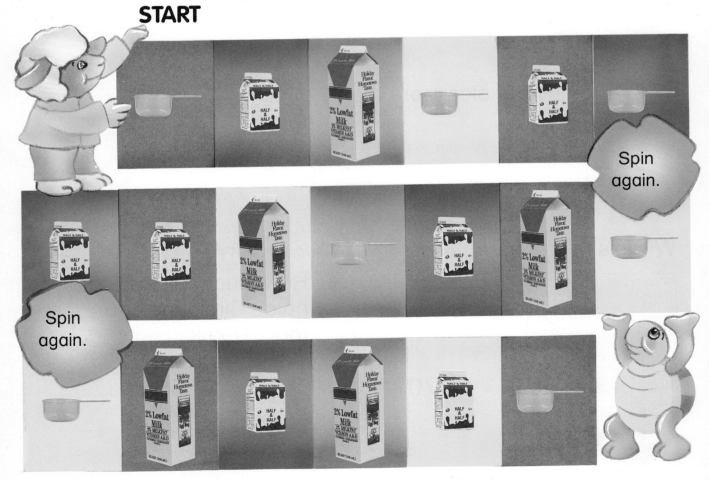

START

Spin again.

Spin again.

END

CHAPTER 29
Adding and Subtracting 2-Digit Numbers

FUN FACTS

A toucan's huge beak is almost the length of its body which is from 13 to 25 inches long.

Theme: In the Rain Forest

Name _____

✓ Check What You Know

Tens and Ones to 100

Write how many tens and ones.
Write the number.

1.

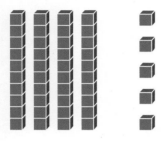

___ tens ___ones = ___

2.

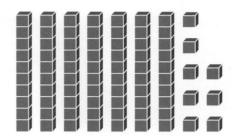

___ tens ___ones = ___

Addition and Subtraction Facts to 20

Add or subtract.

3. 7   4. 9   5. 20   6. 18   7. 10   8. 17
   +8   +7   −10   − 9   + 9   − 9

Fact Families to 20

Write the sum or difference. Circle the two facts if they are in the same fact family.

9. 6 + 9 = ____

16 − 9 = ____

10. 15 − 8 = ____

15 − 7 = ____

11. 9 + 8 = ____

17 − 9 = ____

12. 7 + 6 = ____

14 − 7 = ____

486 four hundred eighty-six   Use this page to review important skills needed for this chapter.

Name _____

Use Mental Math to Add Tens

Learn

A red panda eats 20 leaves.
Then it eats 10 more leaves.
How many leaves does it eat in all?

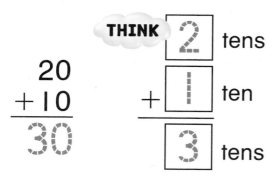

THINK: 2 tens + 1 ten = 3 tens

20
+10

30

It eats __30__ leaves in all.

Start with two tens and add one more ten.

Check

Write how many tens. Then add.

1. 30
    +40

    THINK: ☐ tens + ☐ tens = ☐ tens

2. 50
    +10

    THINK: ☐ tens + ☐ ten = ☐ tens

3. 20
    +70

    THINK: ☐ tens + ☐ tens = ☐ tens

4. 40
    +40

    THINK: ☐ tens + ☐ tens = ☐ tens

Explain It • Daily Reasoning

How does knowing 7 + 2 = 9 help you find the sum for 70 + 20?

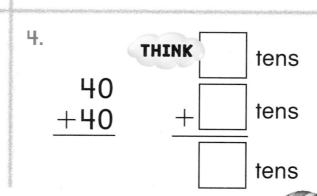

7 + 2 = 9

Chapter 29 • Adding and Subtracting Two-Digit Numbers
four hundred eighty-seven **487**

Practice and Problem Solving

40 + 30 means 4 tens + 3 tens.

THINK

```
  40        4 tens
+ 30      + 3 tens
----      --------
  70        7 tens
```

Add.

1. 50
   +40

2. 30
   +30

3. 50
   +30

4. 10
   +30

5. 20
   +60

6. 10
   +70

7. 20
   +20

8. 80
   +10

9. 10
   +10

10. 70
   +20

11. 10
   +50

12. 20
   +30

13. 40
   +10

14. 30
   +50

15. 60
   +10

Problem Solving

Visual Thinking

Draw what was added.
Complete the number sentence.

16. + =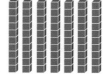

____ + ____ = ____

 Write About It • Look at Exercise 16. Explain how you figured out how many tens to add.

 HOME ACTIVITY • Ask your child to explain how to find the sum for 30 + 20.

Name _____

Add Tens and Ones

HANDS ON Explore

Add.   32
     + 4

STEP 1
Show 32. Show 4.

STEP 2
Add the ones.

STEP 3
Add the tens.

tens	ones
3	2
+	4

tens	ones
3	2
+	4
	6

tens	ones
3	2
+	4
3	6

Connect

Use Workmat 3 and 🎲 to add. Write the sum.

1.
tens	ones
2	5
+	3

2.
tens	ones
4	5
+	2

Explain It • Daily Reasoning

How could you find the sum for 64 + 3 without using blocks?

Chapter 29 • Adding and Subtracting Two-Digit Numbers

Practice and Problem Solving

Use Workmat 3 and to add.
Write the sum.

1.
tens	ones
4	1
+	4
4	5

2.
tens	ones
1	6
+	3

3.
tens	ones
3	5
+	2

4.
tens	ones
2	3
+	3

Problem Solving
Algebra

Write the missing numbers.

5.
tens	ones
3	1
+	☐
3	6

6.
tens	ones
5	3
+	☐
5	9

7.
tens	ones
2	2
+	☐
2	5

8.
tens	ones
6	4
+	☐
6	8

Write About It • What are two ways you could show 19 + 1 with blocks?

HOME ACTIVITY • Ask your child to draw pictures to show how to find the sum for 24 + 5.

Name _____

Add Money

Learn

Add money amounts the same way you add other numbers.

Add the ones. Then add the tens.

Add numbers.   Add money.

```
  23      23       23¢      23¢
+ 15    + 15     + 15¢    + 15¢
----    ----     -----    -----
   8      38       8¢      38¢
```

Check

Add.

1. 32¢
   +27¢
   ____¢

2. 17¢
   +10¢
   ____¢

3. 75¢
   +22¢
   ____¢

4. 43¢
   +16¢
   ____¢

5. 55¢
   +31¢
   ____¢

6. 61¢
   +18¢
   ____¢

7. 24¢
   +33¢
   ____¢

8. 82¢
   + 6¢
   ____¢

9. 10¢
   +29¢
   ____¢

10. 52¢
   + 7¢
   ____¢

11. 53¢
   +15¢
   ____¢

12. 36¢
   +41¢
   ____¢

13. 65¢
   +23¢
   ____¢

14. 24¢
   +50¢
   ____¢

15. 32¢
   +47¢
   ____¢

Explain It • Daily Reasoning

How are pennies like ones?
How are dimes like tens?

Chapter 29 • Adding and Subtracting Two-Digit Numbers   four hundred ninety-one **491**

Practice and Problem Solving

Add.

1. 40¢ + 50¢ = 90¢
2. 72¢ + 23¢ = ___¢
3. 25¢ + 24¢ = ___¢
4. 35¢ + 4¢ = ___¢
5. 19¢ + 20¢ = ___¢

6. 53¢ + 46¢ = ___¢
7. 64¢ + 14¢ = ___¢
8. 75¢ + 3¢ = ___¢
9. 39¢ + 50¢ = ___¢
10. 81¢ + 17¢ = ___¢

11. 44¢ + 33¢ = ___¢
12. 24¢ + 42¢ = ___¢
13. 10¢ + 5¢ = ___¢
14. 61¢ + 8¢ = ___¢
15. 50¢ + 25¢ = ___¢

Problem Solving
Algebra

Write the missing numbers.

16. 4 2 ¢
    +☐ ☐ ¢
    ─────
     6 4 ¢

17. ☐ ☐ ¢
    + 3 3 ¢
    ─────
     6 6 ¢

18. 3 5 ¢
    + ☐ ¢
    ─────
     3 9 ¢

 Write About It • Look at Exercise 18.
Tell how you can subtract to find the missing number.

🏠 **HOME ACTIVITY** • Have your child add a group of 9 or fewer pennies and a group of 9 or fewer dimes.

Name _____

Use Mental Math to Subtract Tens

Learn

A bird finds 40 seeds.
It eats 30 of them.
How many seeds are left?

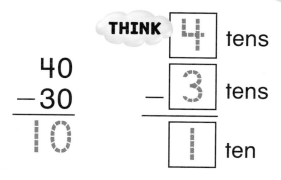

Start with 4 tens and subtract 3 tens.

THINK [4] tens
− [3] tens
   [1] ten

There are __10__ seeds left.

Check

Write how many tens. Then subtract.

1. 90
   −30

   THINK [ ] tens
   − [ ] tens
      [ ] tens

2. 30
   −10

   THINK [ ] tens
   − [ ] ten
      [ ] tens

3. 80
   −40

   THINK [ ] tens
   − [ ] tens
      [ ] tens

4. 70
   −20

   THINK [ ] tens
   − [ ] tens
      [ ] tens

Explain It • Daily Reasoning

How does knowing 6 − 4 = 2 help you find the difference for 60 − 40?

6 − 4   60 − 40

Chapter 29 • Adding and Subtracting Two-Digit Numbers   four hundred ninety-three **493**

Practice and Problem Solving

50 − 20 means
5 tens − 2 tens.

THINK

```
 50      5 tens
-20     -2 tens
---    --------
 30      3 tens
```

Subtract.

1. 70 − 50
2. 40 − 40
3. 60 − 30
4. 80 − 70
5. 40 − 10

6. 90 − 20
7. 50 − 30
8. 20 − 10
9. 60 − 20
10. 70 − 40

11. 80 − 50
12. 30 − 20
13. 90 − 60
14. 50 − 10
15. 40 − 20

Problem Solving
Visual Thinking

Write the number sentence that tells about the picture.

16. ___ ◯ ___ ◯ ___

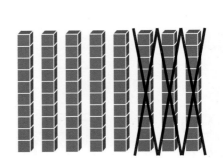

 Write About It • Look at Exercise 16. Write a math story to go with your number sentence.

🏠 HOME ACTIVITY • Ask your child to explain how to find the difference for 70 − 40.

494 four hundred ninety-four

Name _____

Subtract Tens and Ones

Explore

Subtract.   28
          − 5

STEP 1
Show 28.

STEP 2
Subtract the ones.

STEP 3
Subtract the tens.

tens	ones
2	8
−	5

tens	ones
2	8
−	5
	3

tens	ones
2	8
−	5
2	3

Connect

Use Workmat 3 and 🎲 to subtract.
Write the difference.

1.
tens	ones
4	7
−	5

2.
tens	ones
1	9
−	4

Explain It • Daily Reasoning

How could you find the difference for 27 − 4 without using blocks?

Chapter 29 • Adding and Subtracting Two-Digit Numbers   four hundred ninety-five **495**

Practice and Problem Solving

Use Workmat 3 and 🎲 to subtract.
Write the difference.

1. | tens | ones |
    |------|------|
    | 3 | 9 |
    | − | 2 |
    | 3 | 7 |

2. | tens | ones |
    |------|------|
    | 4 | 5 |
    | − | 2 |
    | | |

3. | tens | ones |
    |------|------|
    | 2 | 6 |
    | − | 5 |
    | | |

4. | tens | ones |
    |------|------|
    | 1 | 7 |
    | − | 3 |
    | | |

Problem Solving
Algebra

Write the missing numbers.

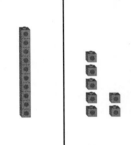

5. | tens | ones |
    |------|------|
    | 4 | 9 |
    | − | ☐ |
    | 4 | 1 |

6. | tens | ones |
    |------|------|
    | 8 | 6 |
    | − | ☐ |
    | 8 | 0 |

7. | tens | ones |
    |------|------|
    | 1 | 8 |
    | − | ☐ |
    | 1 | 4 |

8. | tens | ones |
    |------|------|
    | 7 | 5 |
    | − | ☐ |
    | 7 | 2 |

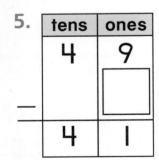

Write About It • Look at Exercise 5. How could you draw 🎲 to prove your answer?

 HOME ACTIVITY • Ask your child to explain how to find the difference for 48 − 3.

Name _____

Subtract Money

Learn

Subtract money amounts the same way you subtract other numbers.

Subtract the ones. Then subtract the tens.

Subtract numbers.   Subtract money.

```
  34        34        34¢       34¢
 -13       -13       -13¢      -13¢
 ---       ---       ----      ----
   1        21         1¢       21¢
```

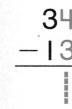

Check

Subtract.

1. 26¢
   −13¢
   ____
     ¢

2. 39¢
   −22¢
   ____
     ¢

3. 68¢
   −32¢
   ____
     ¢

4. 54¢
   −30¢
   ____
     ¢

5. 93¢
   −62¢
   ____
     ¢

6. 69¢
   − 7¢
   ____
     ¢

7. 77¢
   −33¢
   ____
     ¢

8. 97¢
   −76¢
   ____
     ¢

9. 29¢
   −19¢
   ____
     ¢

10. 48¢
   −18¢
   ____
     ¢

11. 86¢
   −23¢
   ____
     ¢

12. 56¢
   −25¢
   ____
     ¢

13. 45¢
   −12¢
   ____
     ¢

14. 69¢
   −45¢
   ____
     ¢

15. 39¢
   −22¢
   ____
     ¢

Explain It • Daily Reasoning

What happens when both numbers have the same ones and tens? Why?

Chapter 29 • Adding and Subtracting Two-Digit Numbers

Practice and Problem Solving

Subtract.

1. 68¢ − 25¢ = 43¢
2. 75¢ − 50¢ = ___¢
3. 89¢ − 64¢ = ___¢
4. 45¢ − 44¢ = ___¢
5. 35¢ − 12¢ = ___¢

6. 56¢ − 36¢ = ___¢
7. 64¢ − 14¢ = ___¢
8. 90¢ − 40¢ = ___¢
9. 28¢ − 24¢ = ___¢
10. 65¢ − 20¢ = ___¢

11. 95¢ − 70¢ = ___¢
12. 83¢ − 22¢ = ___¢
13. 17¢ − 6¢ = ___¢
14. 75¢ − 25¢ = ___¢
15. 88¢ − 28¢ = ___¢

Problem Solving
Application

Write the problem. Solve it.

16. Greg has 49¢.
He spends 24¢ for bird food.
How much does he have left?

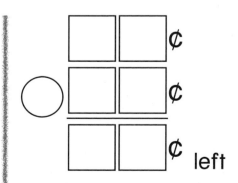

☐☐ ¢
◯ ☐☐ ¢
☐☐ ¢ left

Write About It • Suppose Greg has 68¢ and buys something different. Write your own story problem and solve it.

⬠ **HOME ACTIVITY** • Write a two-digit subtraction problem. Have your child show it with pennies and dimes and solve it.

Name _____

Problem Solving Skill
Make Reasonable Estimates

Without adding or subtracting, circle the best estimate.

THINK:
Using tens can help me estimate.

1. Jon picks 14 flowers.
   Sam picks 5 flowers.
   About how many do they pick in all?

   Too few.   about 2 flowers
              (about 20 flowers)
   Too many. about 200 flowers

2. The school has 30 books about animals.
   Kim takes out 13 books.
   About how many are left?

   about 2 books

   about 20 books

   about 200 books

3. Aman saw 10 monkeys.
   Carla counted 10 more.
   Logan saw 6 other monkeys.
   About how many did they see in all?

   about 10 monkeys

   about 20 monkeys

   about 30 monkeys

4. 40 birds are in the forest.
   12 fly away.
   About how many birds are left in the forest?

   about 5 birds

   about 30 birds

   about 100 birds

Chapter 29 • Adding and Subtracting Two-Digit Numbers

four hundred ninety-nine

Problem Solving Practice

THINK: I do not need an exact answer.

Without adding or subtracting, circle the best estimate.

1. Lana's bird eats 5 seeds for breakfast.
   It eats 5 seeds for lunch.
   It eats 7 more seeds for dinner.
   About how many seeds does it eat in one day?

   about 5 seeds

   about 20 seeds

   about 100 seeds

2. Cam walks 22 steps in the forest.
   She walks 26 more.
   About how many steps does she walk in all?

   about 5 steps

   about 50 steps

   about 500 steps

3. Connor has 50¢.
   He buys a banana for 26¢.
   About how much money does he still have?

   about 5¢

   about 15¢

   about 25¢

4. 100 people are at the park.
   52 of them leave.
   About how many people are still at the park?

   about 50 people

   about 75 people

   about 100 people

HOME ACTIVITY • Ask your child how he or she chose the answer for each problem.

Name _____

Extra Practice

Add or subtract.

1. 30
  +50

2. 70
  +10

3. 20
  +50

4. 90
  −30

5. 60
  −10

Use Workmat 3 and 🔲 to add or subtract.

6.
tens	ones
1	5
+	4

7.
tens	ones
1	9
−	7

Add or subtract.

8. 13¢
  +35¢
  ____¢

9. 20¢
  +47¢
  ____¢

10. 83¢
   −21¢
   ____¢

11. 65¢
   − 3¢
   ____¢

12. 52¢
   −11¢
   ____¢

Problem Solving

Without adding or subtracting, circle the best estimate.

13. Juan sees 12 frogs. Elsa sees 11 more frogs. About how many frogs do they see in all?

about 5 frogs

about 20 frogs

about 100 frogs

Chapter 29 • Adding and Subtracting Two-Digit Numbers

Name _____

✓ Review/Test

Concepts and Skills

Add or subtract.

1. 70
  +20

2. 40
  +30

3. 50
  +40

4. 80
  −10

5. 60
  −20

Use Workmat 3 and ▢ to add or subtract.

6. | tens | ones |
   |------|------|
   | 2 | 4 |
   | + | 5 |
   | | |

7. | tens | ones |
   |------|------|
   | 2 | 8 |
   | − | 4 |
   | | |

Add or subtract.

8. 31¢
  +26¢
  ___¢

9. 15¢
  +32¢
  ___¢

10. 65¢
   −43¢
   ___¢

11. 58¢
   −27¢
   ___¢

12. 26¢
   − 4¢
   ___¢

Problem Solving

Without adding or subtracting, circle the best estimate.

13. Sam sees 25 lizards. Then he sees 12 more. About how many lizards does Sam see?

about 10 lizards

about 40 lizards

about 100 lizards

502 five hundred two

Name _____

★ Standardized Test Prep
Chapters 1–29

Choose the answer for questions 1–4.

1. 53   35   57
   + 4   ○   ○

         45   81
          ○   ○

2. 85¢   15¢   21¢
   −63¢   ○   ○

          22¢   34¢
           ○   ○

3. Which shows 50¢?

○

○

○

○

4. Which is the best estimate?

Rachel jumps rope 23 times. She jumps rope 33 more times. About how many times does she jump rope in all?

about 5   ○
about 15 ○
about 55 ○
about 500 ○

Show What You Know

5. Use . Explain two ways to make 90. Write the numbers.

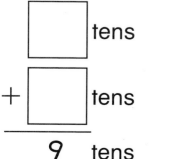

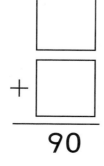

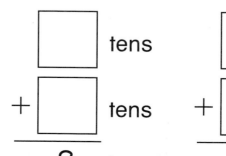

Chapter 29   five hunded three **503**

MATH GAME

Math Path

Play with a partner.

1. Stack the number cards face down.
2. Put your 🎲 on START. Toss the 🎲. Move your 🎲 that many spaces.
3. Take 2 cards. Use the numbers to make a two digit number to complete the problem.
4. Write the problem on paper.
5. Solve. Your partner will check your answer.
6. If you are not correct, lose a turn.
7. The first player to get to END wins.

You will need

2 🎲 🎲

2 sets of 0–6 ⬜

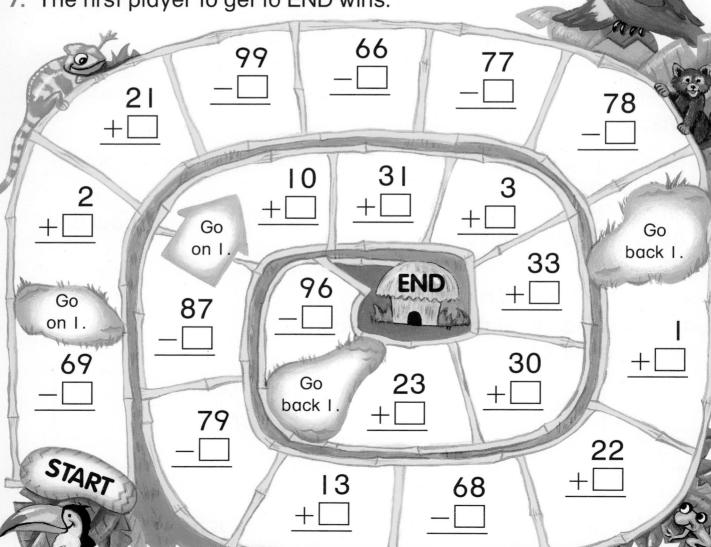

CHAPTER 30 Probability

FUN FACTS

These puppies are red, and black and tan. For every 3 puppies born, 2 are red, and 1 is black and tan.

Theme: Chances

Name _____

✓ Check What You Know

Could It Happen?

1. Circle the picture that shows which is more likely to happen.

2. Circle the picture that shows which is less likely to happen.

Chance

Use a paper clip and a pencil to make a spinner.
Spin 10 times.
Mark a tally mark in the table after each spin.
Circle the color the paper clip landed on more often.

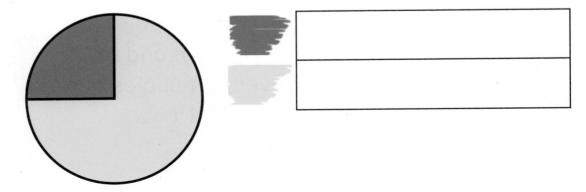

506 five hundred six Use this page to review important skills needed for this chapter.

Name _____

Certain or Impossible

Vocabulary
certain
impossible

Learn

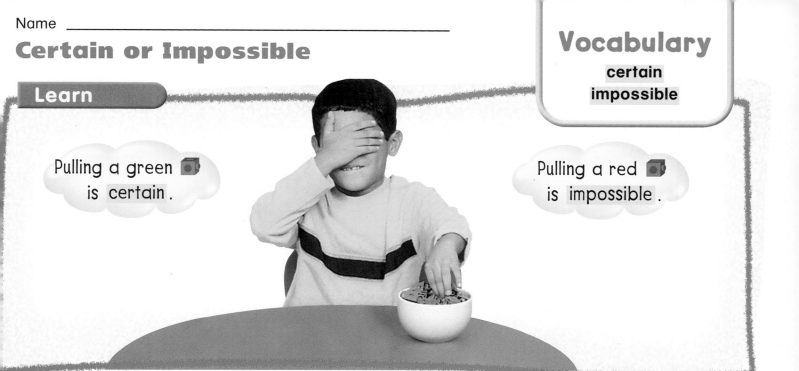

Pulling a green 🟩 is **certain**.

Pulling a red 🟥 is **impossible**.

Check

Mark an X to tell if pulling the cube from the bowl is certain or impossible.

			Certain	Impossible
1.	blue	🥣	X	
2.	red	🥣		
3.	yellow	🥣		
4.	green	🥣		

Explain It • Daily Reasoning

Suppose a bowl had only yellow cubes in it. What color would be certain to be pulled? Explain.

Chapter 30 • Probability

five hundred seven **507**

Practice and Problem Solving

Mark an X to tell if pulling the cube from the bowl is certain or impossible.

			Certain	Impossible
1.	yellow			
2.	red			
3.	green			
4.	blue			

Problem Solving
Application

5. Circle the bowl from which pulling a 🟥 is certain.

Write About It • Draw a bowl from which pulling a 🟥 is certain. Then draw a bowl from which pulling a 🟥 is impossible. Write **certain** or **impossible** under each bowl.

HOME ACTIVITY • Put some pennies in a bowl. Ask your child if pulling a dime from the bowl is certain or impossible. Have him or her explain.

508 five hundred eight

Name _____

More Likely, Less Likely

Vocabulary
more likely
less likely

Learn

Pulling yellow is more likely than pulling green.

Pulling green is less likely than pulling yellow.

Check

Write **more** or **less** to tell how likely each color is to be pulled from the bowl.

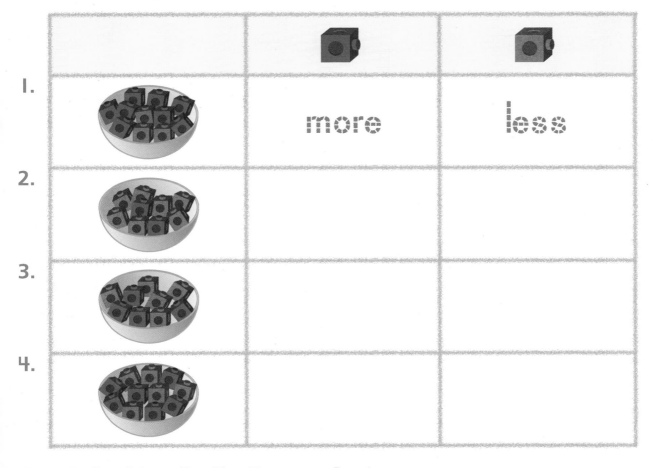

1. more — less

Explain It • Daily Reasoning

How can you tell which color is more likely to be pulled?

Chapter 30 • Probability

Practice and Problem Solving

Draw 🟦 and 🟥 to tell how likely each color is to be pulled from the bowl.

		More Likely	Less Likely
1.	🥣		
2.	🥣		
3.	🥣		
4.	🥣		

Problem Solving
Visual Thinking

5. How could you change the cubes so that pulling a 🟦 is more likely than pulling a 🟥? Draw and color a picture to show your answer.

Write About It • Draw and color a bowl from which pulling a 🟦 is more likely than pulling a 🟥. Then draw and color a bowl from which pulling a 🟥 is more likely than pulling a 🟦.

 HOME ACTIVITY • Put 6 pennies and 3 dimes in a bowl. Ask your child if pulling a dime is more likely or less likely than pulling a penny. Have him or her explain.

510 five hundred ten

Equally Likely

Vocabulary: equally likely

Learn

Pulling yellow and pulling red are **equally likely**.

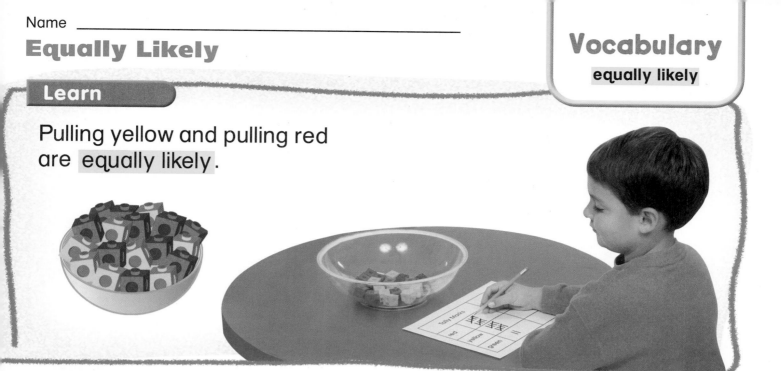

Check

Draw cubes to show which colors are equally likely to be pulled from the bowl.

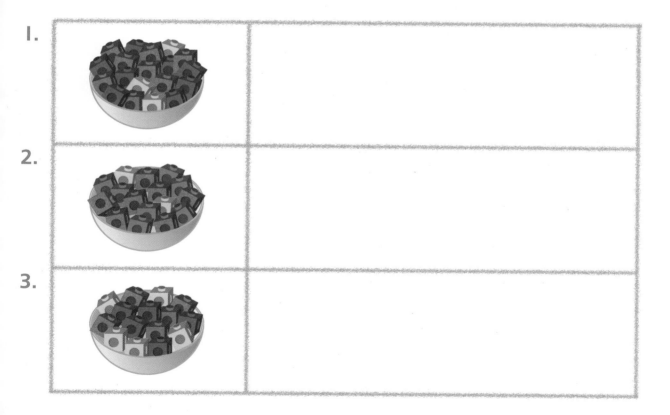

1.
2.
3.

Explain It • Daily Reasoning

What 2 cubes could you add to the last bowl so that pulling 🟥 is less likely than pulling 🟨 or 🟦?

Chapter 30 • Probability

Practice and Problem Solving

Draw cubes to show which colors are equally likely to be pulled from the bowl.

1.

2.

3.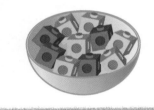

4.

Problem Solving
Visual Thinking

5. How could you change the cubes so that pulling 🎲 and pulling 🎲 are equally likely? Draw and color a picture.

Write About It • Draw and color a bowl from which pulling a 🎲 and pulling a 🎲 are equally likely. Write **equally likely** under the bowl.

 HOME ACTIVITY • Put 6 pennies and 6 dimes in a bowl. Ask your child if a dime is more likely or less likely to be pulled or if both coins are equally likely to be pulled.

512 five hundred twelve

Problem Solving Skill
Make a Prediction

Use a ✏️ and a 📎 to make a spinner.

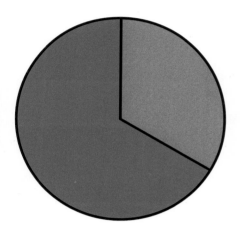

Predict. If you spin the pointer 10 times, on which color will it stop more often?

Circle that color.

red   blue

Check. Spin 10 times.
Make a tally mark after each spin.
Write the totals.

On which color did your pointer stop more often?

	Tally Marks	Total
red		
blue		

1. Predict. If you spin the pointer 10 more times, on which color will it stop more often?
   Circle that color.

   red   blue

   Then spin to check.

Chapter 30 • Probability

five hundred thirteen **513**

Problem Solving Practice

1. Use a ✏️ and a 📎 to make a spinner.

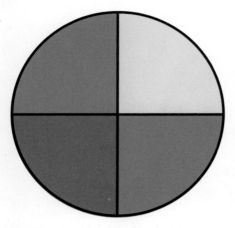

Predict. If you spin the pointer 15 times, on which color will it stop most often?

Circle that color.

green   yellow   blue

Check. Spin 15 times. Make a tally mark after each spin. Write the totals.

On which color did your pointer stop most often?

	Tally Marks	Total
green		
yellow		
blue		

2. Predict. If you spin the pointer 10 more times, on which color will it stop most often?
Circle that color.

green   yellow   blue

Then spin to check.

HOME ACTIVITY • Make a spinner divided into 4 equal parts. Color two parts the same color and each of the other parts a different color. Have your child predict on which color the pointer will land most often. Spin 10 times to check.

Name _____

Extra Practice

Mark an X to tell if pulling the cube from the bowl is certain or impossible.

		Certain	Impossible
1.	blue		

Draw 🟦 and 🟥 to tell how likely each color is to be pulled from the bowl.

		More Likely	Less Likely
2.			

Draw cubes to show which colors are equally likely to be pulled from the bowl.

| 3. | | |

Problem Solving

4. Use a ✏️ and a 📎 to make a spinner. Predict. If you spin the pointer 15 times, on which color will it stop most often? Circle that color.

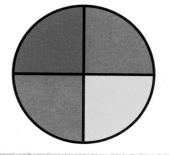

    blue   red   yellow

Check. Spin 15 times. Make a tally mark after each spin. Write the totals.

	Tally Marks	Total
blue		
red		
yellow		

Chapter 30 • Probability

five hundred fifteen **515**

Name _____

✓ Review/Test

Concepts and Skills

Mark an X to tell if pulling the cube from the bowl is certain or impossible.

		Certain	Impossible
1. yellow 🟨 🥣			

Draw 🟥 and ⬛ to tell how likely each color is to be pulled from the bowl.

	More Likely	Less Likely
2. 🥣		

Draw cubes to show which colors are equally likely to be pulled from the bowl.

3. 🥣	

Problem Solving

4. Use a ✏️ and a 📎 to make a spinner. Predict. If you spin the pointer 10 times, on which color will it stop more often? Circle that color.

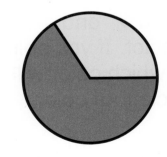

yellow   red

Check. Spin 10 times. Make a tally mark after each spin. Write the totals.

	Tally Marks	Total
yellow		
red		

516   five hundred sixteen

Name _____

★Standardized Test Prep
Chapters 1–30

Choose the answer for questions 1–4.

1. Amanda's mom baked some cookies. Amanda ate 2. Her friend ate 3. How many cookies did they eat altogether?

   8 ○   7 ○   6 ○   5 ○

2. Which completes the sentence?

   It is impossible to pull a _____.

   ○   ○   ○   ○

3. Predict. If you spin the pointer 10 times on which color will it stop more often?

   red ○   green ○

   yellow ○   blue ○

4. Which completes the sentence?

   I am least likely to pull a _____.

   ○   ○   ○

Show What You Know

5.

   Color the cubes to explain which cubes are equally likely and less likely to be pulled from the bowl.

   I am equally likely to pull a ▫ or a ▫.

   I am less likely to pull a ▫ than a ▫.

Chapter 30   five hundred seventeen **517**

IT'S IN THE BAG
Cool Cat Hat

PROJECT You will make a hat and measure with paper clips how far it flies.

You Will Need

- Lunch-size bag
- Paper plate
- Crayons
- Scissors
- Tape
- Paper clips

Directions

1 Cut the circle out of the paper plate. Decorate the hat.

2 Decorate the bag on all sides.

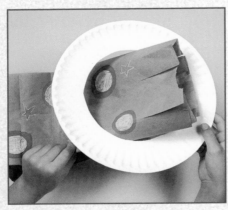

3 Cut slits around the bag. Push them through the paper plate and tape them to it.

4 Throw your hat in the air. Measure with paper clips how far your hat flew.

CAT'S COOL HAT

written by Fay Robinson
illustrated by Lori Lohstoeter

🎩 This book will help me review measurement.

This book belongs to _____.

"Good morning, Turtle," said Cat.
"How do you like my cool hat?"

"It's very nice," said Turtle.
"You do look cool!"

The wind blew Cat's hat into the pond.
"Help!" called Cat.
"I need a long stick to reach my hat."

Turtle said, "Cat, I can get your hat."

Squirrel got a stick.

About how long is this stick?

about ____ 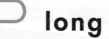 long

Squirrel's stick wasn't long enough.

Turtle said, "Cat, I can get your hat."

Rabbit got a stick.
It wasn't long enough.

About how long is this stick?

about ____ 🖇 long

Cat was not cool now.
"I'll never get my hat back!" he said.

Turtle said, "Cat, I can get your hat."

"Oh, Turtle, I don't think so," said Cat.
"There aren't any sticks long enough."

But Turtle didn't need a stick.
He swam out to the hat.
Then he swam back with
the hat on his back.

"Oh, thank you, Turtle!" said Cat.
"You are the coolest one of all!"

PROBLEM SOLVING ON LOCATION

At the Amusement Park

You can ride two big wooden roller coasters at Camden Park. It is near Huntington, West Virginia.

You can play games to win prizes at the park.

Use the spinners below to decide if your chances of winning are **more likely, less likely** or **equally likely** than your chances of losing. Circle your answer.

roller coaster

1

How likely are you to win if the spinner lands on green?

more likely   less likely   equally likely

2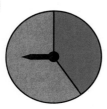

How likely are you to win if the spinner lands on blue?

more likely   less likely   equally likely

3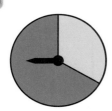

How likely are you to win if the spinner lands on green?

more likely   less likely   equally likely

Unit 6 • Chapters 26–30   five hundred nineteen **519**

CHALLENGE

Area

How many units does it take to cover this shape?

__4__ units

How many units does it take to cover the shape? Use ■. Then count.

1.
   [rectangle]

   _____ units

2.
   [L-shape]

   _____ units

3.
   [rectangle]

   _____ units

4.
   [L-shape]

   _____ units

Name _____

✓ Study Guide and Review

Vocabulary

Draw a line to the tool you use to measure with each unit.

1. **centimeter**

2. **inch**

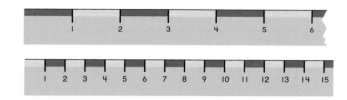

Skills and Concepts

About how long is this?

3. Use an inch ruler to measure. _____ inches

4. Now use a centimeter ruler to measure. _____ centimeters

Look at each object.
Circle the better estimate.

	Object	Estimate
5.	(butter)	about 1 pound about 10 pounds
6.	(oranges)	about 1 pound about 10 pounds

Draw cubes to show which colors are equally likely to be pulled from the bowl.

7.

Read the temperature. Color the thermometer to show the temperature.

8. 20°F

Unit 6 • Study Guide and Review

five hundred twenty-one **521**

Add or subtract.

9. 20
  + 40

10. 60
   − 30

11. 50
   + 30

12. 71¢
   + 23¢
   ―――
      ¢

13. 87¢
   − 25¢
   ―――
      ¢

14. 98¢
   − 50¢
   ―――
      ¢

Use Workmat 3 and 🎲 to add.

15.
tens	ones
2	6
+	3

16.
tens	ones
3	3
+	5

Use Workmat 3 and 🎲 to subtract.

17.
tens	ones
2	9
−	4

18.
tens	ones
4	7
−	3

Problem Solving

Circle the best tool for finding each measurement.

19. Which shoe is heavier?

 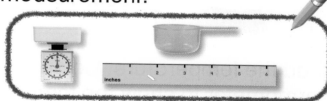

20. Which holds more?

  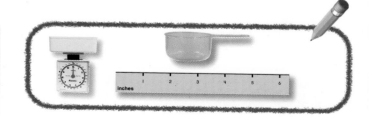

Name _____

✅ Performance Assessment

Hannah's New Lunch Box

Hannah wants to measure her new lunch box.

Here are the tools she can use.

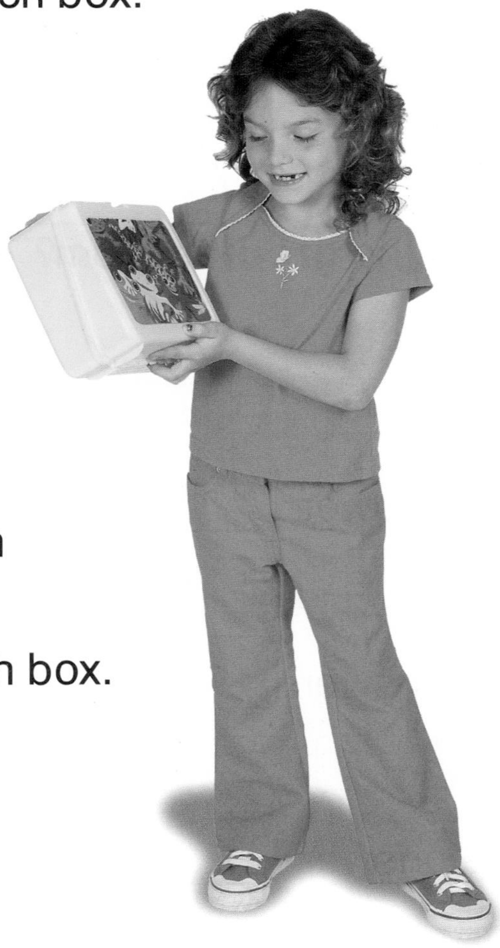

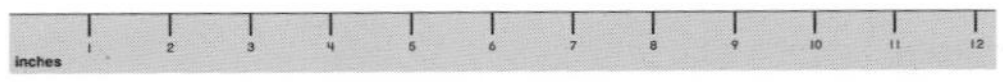

- Circle one of the tools.
- Draw a picture to show how Hannah can use the tool.
- Use this tool to measure a real lunch box.
- Estimate what you think your measurement will be.
- Then measure.

Show your work.

Estimate 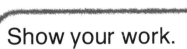   Measurement _____

Name _____

TECHNOLOGY

Calculator • Find the Greatest Sum

$$25 + 35 + 16 = \_\_\_$$
$$65 + 11 + 23 = \_\_\_$$
$$70 + 12 + 14 = \_\_\_$$

Which has the **greatest** sum?

Use a 🖩. Add.

Press [ON/C] [2] [5] [+] [3] [5] [+] [1] [6] [=]

Write the answer. [76]

Press [ON/C] [6] [5] [+] [1] [1] [+] [2] [3] [=]

Write the answer. [ ]

Press [ON/C] [7] [0] [+] [1] [2] [+] [1] [4] [=]

Write the answer. [ ]

Compare. _____ is the greatest sum.

Practice and Problem Solving

Use a 🖩. Find each sum. Circle the greatest sum.

1. $15 + 23 + 31 = \_\_\_$

    $44 + 11 + 13 = \_\_\_$

    $10 + 24 + 65 = \_\_\_$

    Which place value shows which sum is greatest? Underline it.

    tens   ones

2. $21 + 22 + 23 = \_\_\_$

    $51 + 10 + 15 = \_\_\_$

    $45 + 19 + 11 = \_\_\_$

    Which place value shows which sum is greatest? Underline it.

    tens   ones

PICTURE GLOSSARY

above (page 269)

add (page 5)

$$3 + 2 = 5$$

addition sentence (page 9)

$$4 + 1 = 5$$

after (page 181)

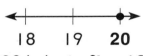

20 is just **after** 19.

afternoon (page 419)

are left (page 33)

3 **are left**.

balance (page 457)

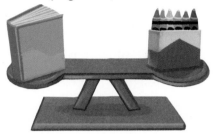

bar graph (page 145)

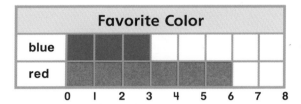

before (page 181)

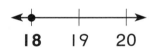

18 is just **before** 19.

below (page 269)

beside (page 270)

between (page 181)

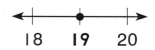

19 is **between** 18 and 20.

centimeter (page 447)

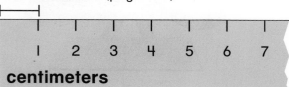

certain (page 507)

Pulling green is **certain**.

chart (page 423)

Subject	Start	End
math		
science		

circle (page 255)

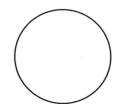

close by (page 269)

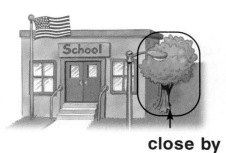

close by

closed figure (page 267)

concrete graph (page 139)

The Fruit Bowl					
apples	🍎	🍎	🍎		
oranges	🍊	🍊			
bananas	🍌	🍌	🍌	🍌	

cone (page 251)

526 five hundred twenty-six

count back (page 101)

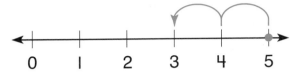

$5 - 2 = 3$

Start at 5. **Count back** two.
You are on 3.

count backward (page 183)

52, 51, 50

Count backward from 52.

count forward (page 183)

52, 53, 54

Count forward from 52.

count on (page 71)

$8 + 2 = 10$

Say 8. **Count on** two.
9, 10

cube (page 251)

cup (page 473)

cylinder (page 251)

day (page 417)

The **days** of the week are:
Sunday, Monday, Tuesday, Wednesday, Thursday, Friday, and **Saturday.**

difference (page 35)

$9 - 3 = 6$ ← difference

dime (page 369)

or 10¢
10 cents

dollar (page 387)

1 dollar = 100¢

doubles (page 75)

$4 + 4 = 8$

doubles plus one (page 213)

$4 + 4 = 8$, so $4 + 5 = 9$

down (page 271)

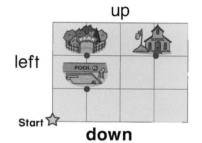

equal parts (page 351)

equally likely (page 511)

Pulling blue and red are **equally likely**.

equals = (page 5)
the same as

4 + 1 = 5

4 plus 1 **equals** 5.

estimate (page 167)

about 10 buttons

even numbers (page 199)

0, 2, 4, 6, 8, 10 . . .

evening (page 419)

face (page 253)

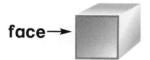

fact family (page 121)

5 + 3 = 8   3 + 5 = 8

8 − 3 = 5   8 − 5 = 3

far (page 269)

feet (page 445)

Use **feet** to measure longer objects.

fewest (page 384)

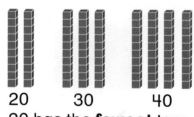

20   30   40

20 has the **fewest** tens.

foot (page 445)

12 inches = 1 **foot**

gram (page 461)

This paper clip is about 1 **gram**.

half dollar (page 387)

50¢ or half dollar

half hour (page 407)

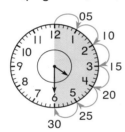

4:30

There are 30 minutes in a **half hour**.

hour (page 405)

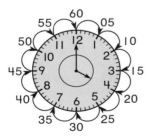

4:00

There are 60 minutes in an **hour**.

hour hand (page 401)

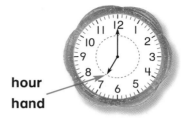

hour hand

hundred (page 163)

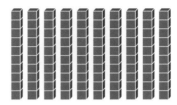

impossible (page 507)

Pulling red is **impossible**.

in all (page 3)

There are 3 **in all**.

inch (page 443)

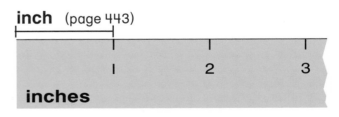

inches

is equal to (page 180)

25 **is equal to** 25.

25 = 25

is greater than (page 175)

5 **is greater than** 1.

5 > 1

is less than (page 177)

3 **is less than** 5.

3 < 5

kilogram (page 461)

This large book is about 1 **kilogram**.

left (page 271)

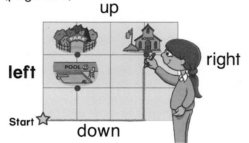

less likely (page 509)

Pulling red is **less likely** than pulling blue.

line of symmetry (page 273)

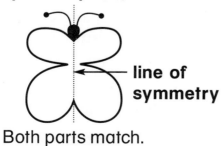

Both parts match.

liter (page 475)

longest (page 439)

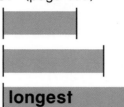

make a ten (page 303)

Move 1 counter into the ten frame. **Make a ten.**

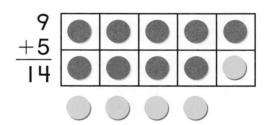

minus (page 35)

3 − 2 = 1

3 **minus** 2 equals 1.

minute (page 403)

You can estimate a **minute**.

minute hand (page 401)

530 five hundred thirty